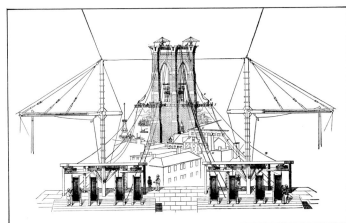 

A PICTURE HISTORY OF THE
BROOKLYN BRIDGE

With 167 Prints and Photographs

Mary J. Shapiro

DOVER PUBLICATIONS, INC., New York

ACKNOWLEDGMENTS

It would be difficult indeed to write anything about the Brooklyn Bridge without referring to David McCullough's exhaustive and highly acclaimed book *The Great Bridge: The Epic Story of the Building of the Brooklyn Bridge*. His superb and thoroughly documented study was of inestimable value in guiding my research and selection of prints and photographs.

I am also deeply indebted to the many archivists and librarians upon whose expert and cheerful assistance I greatly relied. In particular I owe much to: the staffs of the Museum of the City of New York, the New-York Historical Society, the New York Public Library; Elizabeth Stewart, Head of Special Collections and Frieda Gray, Assistant Archivist of the Library of Rensselaer Polytechnic Institute, Troy, New York; Clark Beck, Assistant Curator of the Special Collections Department of the Alexander Library at Rutgers University, New Brunswick, New Jersey; Patricia Flavin, Acting Director of the Long Island Historical Society; Norman Brouwer of the South Street Seaport Museum; and David Wooters of George Eastman House in Rochester, New York.

Blair Birdsall, who was Chief Engineer of John A. Roebling's Sons Company from 1934 to 1964, very generously allowed me to select many photographs and prints from his own personal collection. Also, Hugh Dunne of the Metropolitan Transit Authority made available some very crucial pictorial material as did Steve Gera of the New York City Department of Transportation. I would also like to thank Paul O'Dwyer, Mary Herrington, Ralph Greenhill and Robert Vogel for their advice and help.

I am especially grateful to Hayward Cirker, whose idea it was to do this book, and to my editor, James Spero, for his invaluable advice and expertise.

To thank my husband Barry and sons Michael and Eben for their long-suffering forbearance and greatly needed assistance, would be to acknowledge only the least of my debts to them.

For Margaret and Peter J. Crotty
with love.

Published in Canada by General Publishing Company, Ltd., 30 Lesmill Road, Don Mills, Toronto, Ontario.
Published in the United Kingdom by Constable and Company, Ltd., 10 Orange Street, London WC2H 7EG.

A Picture History of the Brooklyn Bridge: With 167 Prints and Photographs is a new work, first published by Dover Publications, Inc., in 1983.

Book design by Carol Belanger Grafton

Manufactured in the United States of America
Dover Publications, Inc.
180 Varick Street
New York, N.Y. 10014

Library of Congress Cataloging in Publication Data

Shapiro, Mary J.
A picture history of the Brooklyn Bridge.

1. Brooklyn Bridge (New York, N.Y.)—History.   I. Title.
TG25.N53S5   1983   388.1′32′097471   82-9506
ISBN 0-486-24403-2   AACR2

INTRODUCTION

"All that the age had just cause for pride in—its advances in science, its skill in handling iron, its personal heroism in the face of dangerous industrial processes, its willingness to attempt the untried and the impossible—came to a head in Brooklyn Bridge."

Lewis Mumford, 1924.

Today, in an age when we can land a man on the moon and explore the outermost planets of the solar system by satellite, it is difficult to imagine the impact that the completion of the Brooklyn Bridge had just 100 years ago. For its time, it was an engineering achievement of almost miraculous proportions. Its long sweep of steel, etching a gentle arch across the sky, created a futuristic backdrop for nineteenth-century New York, where most buildings were a mere four or five stories high and the chief means of transportation was the horse and buggy. It was by far the largest single-span suspension bridge in the world, a technological giant step in the development of suspended structures.

The suspension bridge has been in man's building repertoire for thousands of years. Primitive suspension bridges consisted of two parallel ropes supporting a wooden walkway fastened at either end to a rock, tree or whatever else happened to be available. The first modern suspension bridge, however, was comparatively recent. Consisting of a wooden-plank roadway suspended from cables of iron chains, it was invented by James Finley, a Pennsylvania preacher, in 1796. By 1810, 50 bridges with hand-forged chains had been built according to his patent. Finley's first bridge had a span of only 70', but one of his later bridges was 306' long, a considerable achievement.

In 1825, the first great suspension bridge was built by Thomas Telford across the Menai Straits in Wales. The main span, suspended from iron chains, was 579' from tower to tower. In 1834, the Grand Pont at Fribourg, Switzerland, a suspension bridge with a main span of 870', was opened. In 1849, Charles Ellet completed his Wheeling Bridge in West Virginia, a single span of 1,010'.

This rapid development of the modern suspension bridge during the nineteenth century was unfortunately marred by some spectacular failures due to mistaken theories and simple lack of technical information. With disturbing regularity, suspended structures collapsed under loads they were calculated to support or were blown apart by winds they were designed to withstand.

In 1839, a raging wind and rain storm tore up the deck of Thomas Telford's Menai Straits Bridge. The bridgekeeper had to row across the straits in the middle of the night to stop the driver of the London mail coach from galloping to his death. In 1830, over 200 people were watching a boat race from the deck of a suspension bridge in Montrose, England. As they ran from one side to the other, one of the chains broke, causing the platform to plunge into the river. Many of the people drowned. In 1850, in Angers, France, a whole battalion of marching soldiers were drowned when a suspension bridge collapsed beneath them. In 1854, Charles Ellet's Wheeling Bridge collapsed in a gale, just five years after it was completed. A newspaper report of the failure described the rolling undulation of the bridge floor and the momentum of the thrashing deck finally wrenching loose from its cables. The picture created is very much like a wood engraving of the Bridge of Constantine across the Seine, which collapsed in a high wind in 1872. And a modern disaster occurred in 1940 when the Tacoma Narrows Bridge, a span of 2,800' dubbed "Galloping Gertie," plunged into Puget Sound in a wind of only 42 mph. (The Brooklyn Bridge has withstood winds of over 70 mph.)

FRANCE.—THE ACCIDENT AT THE BRIDGE OF CONSTANTINE—THE ROAD-BED OF THE BRIDGE

During the construction of the Brooklyn Bridge there was understandable apprehension expressed in the newspapers concerning the stability and safety of a suspension bridge. The findings of "experts" were published attesting to the inadvisability of raising so precarious a structure in such turbulent waters, over such a long distance. It is a tribute to the designers and builders of the bridge that they were able to overcome these obstacles and build an absolutely stable structure that endures today.

The Brooklyn Bridge was designed by John A. Roebling, a German immigrant who received his degree in civil engineering from the Royal Polytechnic Institute in Berlin in 1826. His special interest was the design of suspension bridges, which was the subject of his graduation thesis. Roebling came to the United States in 1831, but it was not until 1844 that he received his first chance to build a suspended structure. It was a modest aqueduct but strong enough to carry 2,100 tons of water. From 1845 until his death in 1869, Roebling designed five major suspension bridges: the Monongahela Bridge in Pittsburgh, 1845; the Niagara Bridge, 1855; the Allegheny Bridge in Pittsburgh, 1860; the Cincinnati-Covington Bridge, 1867; and the Brooklyn Bridge, 1883. Of these, the Cincinnati-Covington Bridge and the Brooklyn Bridge are the only two to survive today.

Brooklyn Bridge was to be the crowning achievement of Roebling's career, but in 1869, before construction was even underway, he was killed in a freakish accident while siting the position of the Brooklyn tower. The enormous task of making his ambitious design a reality was left to his son, Washington A. Roebling.

The younger Roebling had worked with his father on the Allegheny and Cincinnati-Covington Bridges and was thoroughly acquainted with the plans and designs for the Brooklyn Bridge. He was 32 years old when he took his father's place as Chief Engineer. Tragically, three years later, he suffered a severe attack of the bends while working in a pneumatic caisson beneath the New York tower, 78' below high water. The attack rendered him an invalid for the rest of his life.

Despite the great suffering he had to endure, Roebling managed to oversee every detail of the construction work from his home in Brooklyn Heights. The sheer immensity of the project—the massive weights and giant dimensions—presented tremendous logistical problems which he had to solve step by step. As Roebling would note after the bridge was completed, "There is scarcely a feature in the whole work that did not present new and untried problems."

The total length of the bridge is 5,989' and the central span is 1,595'6"—at that time, the longest single span ever constructed, over one third longer than Roebling's Cincinnati-Covington Bridge. The roadway is held aloft by four massive cables, each 15¾" thick and 3,578'6" long. The cables rest on the towers, forming a natural catenary or curve over the river. Each tower is 276'6" high. The cables are fastened on either side of the river to wrought-iron anchor plates buried beneath granite anchorages which weigh 60,000 tons each. (There are four plates per anchorage, each plate weighing 23 tons.) The anchorages are 89' high and measure 129' × 119' at their bases. Suspenders attached to the cables hold up the 85'-wide floor, which is 5' wider than Broadway. Roebling had divided the roadway into five lanes, the outermost for horses and carriages, the intermediate for a rapid-transit system of cable cars which would cost five cents per ride, and a central raised promenade which remains today as it was when the bridge first opened.

Brooklyn Bridge was the first suspension bridge to be constructed of steel, a stronger and lighter material than iron. The steel wire was galvanized with zinc to protect it from rust. The floor of the bridge was also made of steel and reinforced beyond the original specifications to accommodate heavy locomotives. There was a tentative plan to run the New York Central over the bridge to terminate in Brooklyn. This never happened but the strengthened floor insured that the bridge would survive the radical transition from horse and buggy to automobile traffic, a change that could not possibly have been foreseen when the bridge was designed. Brooklyn Bridge was also the first such structure to be illuminated by electric lights.

The bridge was completed in 1883, after 14 years of construc-

tion, a technological odyssey through a nineteenth-century political morass of graft and greed. It was the age of William M. Tweed, who had considerable control in all major New York building projects until his arrest in 1871. At his trial Tweed admitted that Brooklyn State Senator Henry Murphy had given him $60,000 to influence members of the New York City Common Council to purchase stock from the newly incorporated New York Bridge Company, a private concern. In return for this favor, Tweed became a major stockholder in the company and a member of its board of trustees.

After Tweed's fall from power, a new charter was passed in Albany making the bridge a public property to be managed by the Mayor and Comptroller of Brooklyn and New York and their appointees. Despite the spirit of reform, the bridge project was plagued by accusations and investigative committees. The worst fraud, one with potentially serious consequences, occurred in 1878, when Roebling's engineers discovered that J. Lloyd Haigh, who was under contract to provide steel wire for the bridge cables, was smuggling substandard wire onto the bridge and that the condemned wire had been irrevocably woven into the cables along with the good wire.

When the fraud was made public, Henry George, writing in *Frank Leslie's Illustrated News*, remarked on the irony of achievement in the Brooklyn Bridge—a work of technical excellence assailed by gross political corruption: "The East River Bridge is a crowning triumph of mechanical skill; but to get it built, a leading citizen of Brooklyn had to carry to New York sixty thousand dollars in a carpetbag to bribe a New York alderman. The human soul that thought out the great bridge is prisoned in a crazed and broken body that lies bed-fast, and could only watch it grow by peering through a telescope. Nevertheless, the weight of the immense mass is estimated and adjusted for every inch. But the skill of the engineer could not prevent condemned wire from being smuggled into the cable."

The Brooklyn Bridge was officially opened on May 24, 1883 and the great occasion was marked by uproarious exuberance as thousands of sightseers crowded both cities vying for the opportunity to be first across the new bridge. The President of the United States, Chester A. Arthur, accompanied by Grover Cleveland, Governor of New York State, Franklin Edson, the Mayor of New York, and Seth Low, the Mayor of Brooklyn, presided over the opening ceremonies. The day's activities included several hours of speeches in the afternoon, a reception at Washington Roebling's house in Brooklyn Heights, a grand dinner at Seth Low's home and, in the evening, an hour-long fireworks display which culminated in the simultaneous explosion of 500 monster rockets. This was followed by a reception at the Brooklyn Academy of Music.

The Brooklyn Bridge has been celebrated ever since by artists, poets, art critics, social historians, photographers, politicians, writers and anyone who walks or drives its length, is attracted by its magical "flight of strings"[1] or responds to its presence as a constant source of joy and inspiration.

This collection of photographs, drawings and wood engravings, selected from a great variety of sources, including private collections, historical societies, museums, libraries, universities and city archives, offers its own celebration and endeavors to document and illustrate the creation and enduring presence of the greatest engineering achievement of the nineteenth century: the Brooklyn Bridge.

[1]Hart Crane, *Proem: To Brooklyn Bridge.*

A PICTURE HISTORY OF THE
BROOKLYN BRIDGE

1. The Brooklyn Ferry in 1750, from the Brooklyn shore. The settlement of Breukelen (meaning marshland), as it was originally called, was established by a group of Dutch farmers in 1646. But the first regular ferry service to Manhattan was not started until 100 years later by an enterprising citizen of New Amsterdam named Cornelius Dirckson. Sail and rowboats were used for people, and scows with sprit sails (as shown in this wood engraving) for livestock and wagons of produce from Long Island's rich farmlands. Passage was rather dangerous, and it was not unusual for a boat to capsize or be caught in a sudden strong wind, often with fatal consequences. (The Brooklyn tower of the bridge would be built just to the right of the small houses depicted in this illustration.) **2. Site of the Brooklyn Bridge approach, ca. 1820.** Francis Guy, who lived in Brooklyn from 1817 to 1820, painted this charming scene, showing the intersection of Dock and Front Streets, from the window of his home at 11 Front Street. James Street, which is now covered by the approach to the bridge, cuts diagonally back to the center of the painting. Brooklyn at this time was still a remote, picturesque village far from the frenzied activity of Manhattan. In 1810 it had a population of about 3,000 people. But by 1850 Brooklyn's bound-aries had spread to include 25 surrounding villages, with a population of almost 100,000. Over one tenth of these people worked in Manhattan and thus had to commute twice daily on one of the several East River ferries. **3 (Overleaf). The Fulton Ferry from the New York shore, ca. 1847.** In 1814 the Fulton Ferry was established, using Robert Fulton's steam-powered ferryboats. It left from the foot of Fulton Street in Manhattan and traveled upstream to the foot of Fulton Street in Brooklyn. In 1849, shortly after this lithograph was published, the *New York Tribune* wrote: "Ferries are rapidly becoming unequal to the immense and swiftly increasing intercourse between counting house and home to so many thousands of our citizens. The only thing to be thought of is a bridge . . . affording passage to a steady stream of vehicles and pedestrians." Twenty years later, when construction of the Brooklyn Bridge had just started, the six ferry lines between New York and Brooklyn carried 110,000 passengers daily. This view shows Manhattan in the foreground with the Fulton Market on the left and Schermerhorn Row on the right. Across and upriver is the Brooklyn terminus at the foot of Fulton Street. Just to the left is the site of the future Brooklyn tower of the bridge.

1

2

3

FULTON FERRY TO BROOKLYN

4

5

6

7

4. Thomas Pope's Rainbow Bridge, 1811. Talk of a bridge connecting Brooklyn and New York had started as early as 1800, when Jeremiah Johnson (mayor of Brooklyn from 1837 to 1840) entered in his diary that such a bridge would "find more weight in the practicality of the scheme than at first view imagined." He knew of a man who could build a bridge in two years' time. That man may have been Thomas Pope, who in 1811 published *A Treatise on Bridge Architecture*. Pope wrote that he had devised a special method of building a wooden cantilever bridge which could span great distances. He designed his bridge for the Hudson River, proposing a 1,800′ wooden structure which would soar 223′ above high water. (The Brooklyn Bridge is 135′ above high water.) The drawing of his "flying pendent lever bridge," although impractical and untried, was, at that time, a beautiful and daring vision, like "a rainbow rising on the shore." Pope accompanied his treatise throughout with heroic couplets, which he composed himself, an example of which appears under this engraving:

> Let the Broad Arch the spacious Hudson stride,
> And span Columbias Rivers far more wide,
> Convince the World America begins,
> To foster Arts the ancient work of Kings.

5. Ice bridge over the East River, 1867. The winter of 1866–67 was particularly severe; for weeks the East River was clogged with great chunks of ice which kept the ferries from running. As *Harper's Monthly* remarked, there were passengers who could travel from New York to Albany and "arrive earlier than those who set out the same morning from their breakfast tables in Brooklyn for their desks in New York." It was during this period that public agitation for a bridge began to be heard, making it possible for a handful of Brooklyn businessmen to create a coalition of various interests in order to get the project underway. This illustration shows a continu-

ous sheet of ice which was formed shore to shore on January 23, 1867. On this occasion, thousands of people attempted to cross the river on foot—a very hazardous undertaking, since the ice immediately broke up when the tide turned, leaving many stranded on floating chunks of ice. However, few fatalities were reported, thanks to the rescue efforts of a variety of vessels, tugboats and ferries.

6. William C. Kingsley (1833–1885). A Brooklyn contractor, Kingsley had been promoting the idea of a bridge since 1865 when, at 32, he employed an engineer, and fellow Brooklyner, Col. Julius Adams, to draw up a design with cost estimates for material and labor. Kingsley was very active in Brooklyn politics and had worked on several important public-works projects, including Prospect Park and the Hempstead Reservoir. His company, Kingsley & Keeney Contractors, had made him worth close to a million dollars.

7. Henry Cruse Murphy (1810–1882). State Senator Henry Cruse Murphy, a respected Brooklyn lawyer who had been mayor of Brooklyn and was later American Minister to The Hague under President Buchanan, became interested in the bridge project after a meeting with Kingsley, who convinced him of the urgency of having a bill passed in Albany chartering the New York Bridge Company. Murphy drafted the incorporating charter and submitted it to the state legislature on January 25, 1867. It passed on April 16, but it would be another two years before actual work on the bridge was started. Murphy spent a major part of his adult life as president of the New York Bridge Company, from 1867 until his death in 1882. Like Moses, he led Brooklyn to the promised land and yet was denied the satisfaction of ever entering himself. Six months before the Brooklyn Bridge was officially opened, Murphy died at the age of 72, after successfully guiding the project through a bureaucratic labyrinth of political snarls and squabbles.

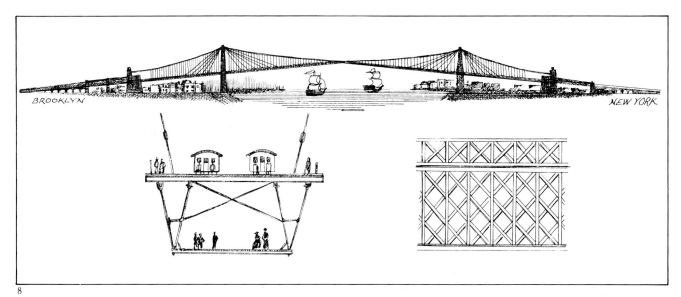

8. Bridge designed by Julius Adams, 1865. An early drawing of Adams' design for a bridge, done for Kingsley, was published in the *Union*, a Brooklyn newspaper, on February 8, 1866. The accompanying article stated that the bridge would run from a point near the head of Main Street in Brooklyn to Catherine Street between Madison and Henry Streets in New York, would be 4,250′ long with a central span of 1,350′ and would cost approximately $3 million. The *Union* worried about the feasibility of the plan and was especially concerned about the difficulty of fixing guy wires to stabilize the bridge. Noting that no such guys are shown in the design, the paper comments trustingly: "He has probably left them out as involving too much intricacy of detail in the drawing and not necessary to the comprehension of his plan. Without guys . . . the bridge would be blown away by the first breeze strong enough to drive a ship." Henry Murphy took a drawing of Adams' bridge to Albany with him and passed out copies to fellow legislators to drum up support for the project. Adams' price tag was an especially attractive, albeit unrealistic, feature. **9. John Augustus Roebling (1806–1869).** Born in Mühlhausen, Germany, Roebling had studied engineering at the prestigious Royal Polytechnic Institute in Berlin, where he received his degree in civil engineering in 1826. Here he also studied philosophy under Hegel, who, it is said, considered Roebling his favorite student. Roebling was undoubtedly a genius with an extraordinary sense of purpose and resolve who acted on Hegel's belief that America "was the land of desire for all those who are weary of the historical lumber room of old Europe." As Roebling himself wrote in his diary, "It is not contempt for our Fatherland that causes us to leave it . . . it was an inclination and an ardent desire that our circumstances may be bettered." This photo was made shortly before Roebling died at the age of 63. **10–12. Pages from John A. Roebling's Notebooks on Bridge Building, 1824/25.** At the Royal Polytechnic Institute, J. F. W. Dietlein introduced his students to the principles of suspension-bridge construction and described in detail current examples of this new form of engineering. Roebling's very precise notes on Dietlein's lectures include drawings of a 210′ suspended span which recently had been raised over the Main River near Bamberg, Germany. It was the first such structure built in Germany and, as the legend goes, inspired Roebling in his own lifetime ambition to build suspension bridges. After his graduation, Roebling worked for three years as an assistant engineer for the Prussian government, building roads and working on other public projects. Frustrated by a restrictive bureaucracy that stymied all creative and innovative efforts, Roebling gave up his apprenticeship and returned to his home in Mühlhausen in 1830. There he organized a small group of men and women willing to make the difficult journey to America to start a new life.

Zweiter Theil

des

Brückenbaues,

den

Bau der steinernen Brücken enthaltend.

Vorgetragen von d. H. Dctr: Dietlein

im Wintersemester 1824/25.

Ausgearb: v. A. Röbling

11

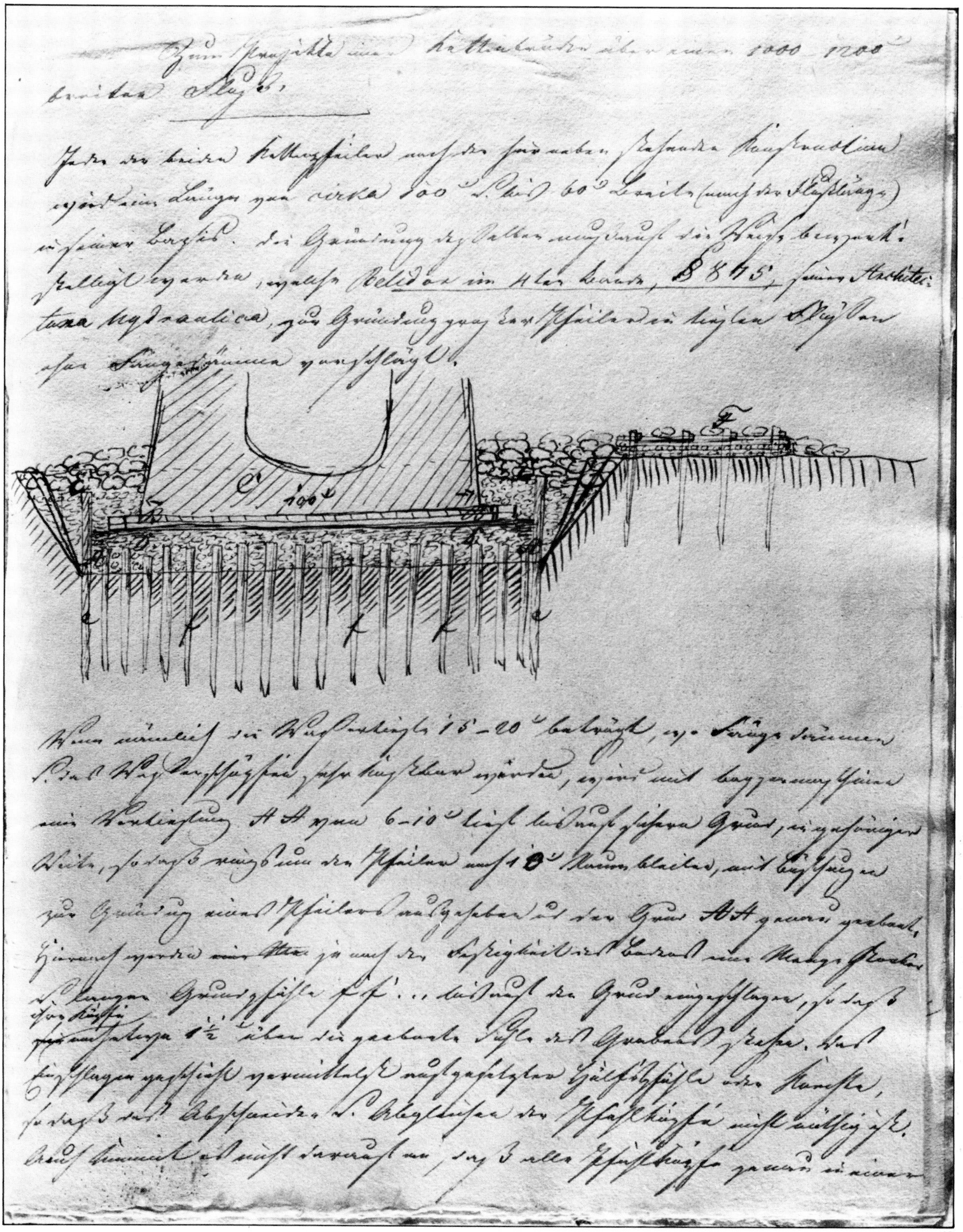

12

Nach der Natur gezeichnet von T. Gosewisch im Juli 1835.   Steindruck von E. W. Robling in Mühlhausen

Sachsenburg,
Colonie von Thüringern und Sachsen bei Pittsburg.
(Von der Südseite.)

12 Herling 10 Mader 9 Tolle 8 Berreigau 7 Hübgen Schmied Mühl Dunkelroth Leonau Lamp Monscher 2 Bär   1 Geb. Robling
Schmied 11 Reichardt Kaffeefarm bei Seiberg   Tischler   Sch. Wirthschaft Feld Fuhrmann und Trappen Kung und Trappen

13

14

12

15

13. Saxonburg, 1835. Roebling came to the United States in 1831 with his brother Karl and a group of German farmers and tradesmen bent on establishing a utopian community in the New World. They bought land in western Pennsylvania just north of Pittsburgh and laid out a town which they called Saxonburg (Sachsenburg). In 1836 Roebling married the daughter of another Saxonburg settler, and together they had six children, the oldest of whom was Washington Augustus Roebling (born 1837), who would eventually become his father's most important assistant and still later take his place as Chief Engineer of the Brooklyn Bridge. **14. The Pennsylvania Portage Railroad.** During his stay in this small community, Roebling worked as a civil engineer for the state of Pennsylvania. While he was engaged in surveying a new railroad route, Roebling observed the canalboat inclined railway (portage railroad) between Hollidaysburg and Johnstown which was part of the Pennsylvania canal route from Philadelphia and Pittsburgh. Barges had to be placed on tracks and hauled by cable over the Allegheny Mountains from one canal to another. Hemp ropes 6″ in diameter were used for this operation, but they often broke, causing delays, accidents and added expense. In a small shed behind his home in Saxonburg, Roebling devised a method for twisting strands of wire into an iron rope which was far less cumbersome and much more durable than hemp. Roebling was eventually given a contract to produce this new cable for the entire Pennsylvania Canal. This was the beginning of his wire-rope business which he moved to Trenton, New Jersey in 1848. **15. Delaware Aqueduct, 1848.** In 1844, Roebling received a contract to build an aqueduct over the Allegheny River at Pittsburgh. This was his first chance to build a suspended structure. Using his own wire rope, he built a seven-span aqueduct capable of carrying over 2,100 tons of water. It was pulled down in 1861, but the Delaware Aqueduct over the Delaware River at Lackawaxen, Pennsylvania, was built in 1848 and still stands today. It has four spans varying in length from 131′ to 142′, and is one of four aqueducts Roebling built on the Delaware and Hudson Canal between 1847 and 1850.

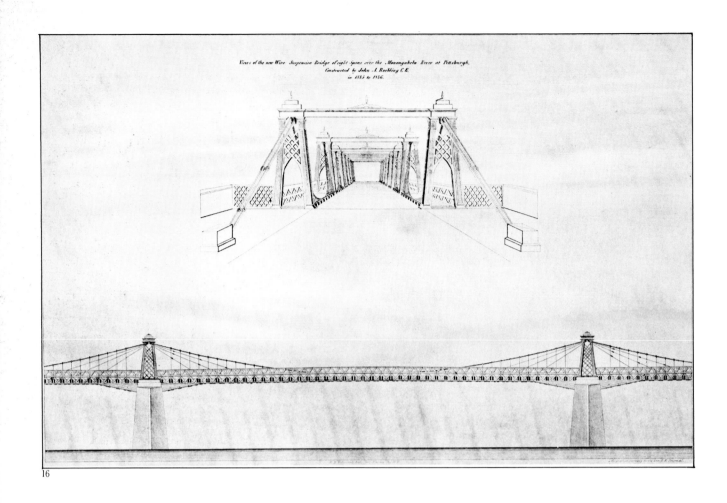

Views of the new Wire Suspension Bridge of eight Spans over the Monongahela River at Pittsburgh.
Constructed by John A. Roebling C.E.
in 1845 to 1846.

16

16. The Monongahela Bridge, 1845. The Monongahela Bridge in Pittsburgh was the first suspension bridge designed and built by John Roebling. It was started in 1845 and completed by the next year. It had eight spans, each 188′ long, suspended from cast-iron towers. Roebling used his wire rope, spinning the cables in place over the supporting towers just as he had on the aqueduct which was completed the year before. He also used iron bars imbedded in heavy concrete blocks to anchor the wire rope. The inclined stays of iron bars made the bridge stiffer and more stable. In his later bridges Roebling would use iron rope for this purpose. ***17 & 18.* Two views of the Niagara Bridge, 1851–1855.** Roebling began the awesome task of spanning the Niagara River with a railroad suspension bridge in 1851. It was a single span of 821′4″ with two levels suspended from separate cables. The tracks of the Great Western Railroad of Canada ran across the top level while the bottom was used for carriage and foot traffic. The bridge was further braced by a system of stays beneath the bridge floor. Whereas Roebling concluded that there were "no vibrations whatever" and proudly wrote his family that "no one is afraid to cross," there was at least one person who had a different opinion. Mark Twain observed: "You drive over to Suspension Bridge and divide your misery between the chances of smashing down two hundred feet into the river below, and the chances of having a railway-train overhead smash down onto you."

17

18

19

20

21

22

19. Wheeling Suspension Bridge, 1849. The Wheeling Bridge, a span of 1,010′, was the longest suspension bridge yet attempted. Built by Charles Ellet in 1849, it collapsed in a storm in 1854, just five years after its completion. Roebling, who at that time was engaged in building his Niagara Bridge, decided that Ellet's bridge could not withstand the tremendous winds to which it was exposed because of the lack of stiffness of the bridge floor. Roebling said the bridge "was destroyed by the momentum acquired of its own dead weight, when swayed up and down by the wind." To make the floor of his own Niagara Bridge even stiffer, Roebling strengthened the trusswork which runs longitudinally beneath the bridge floor.

20. Allegheny Suspension Bridge, 1858–1860. The Allegheny suspension bridge, in Pittsburgh, Pennsylvania, was a multispan structure 1,037′5″ long, designed for heavy road travel. Its floor was 40′ wide and was suspended from 45′-high cast-iron columns with ornamental caps. Roebling designed the ornamental towerwork, which was much praised at the time; as Roebling himself wrote home, "The bridge will be beautiful." The stiffness of the bridge floor was insured by two longitudinal iron lattice girders located beneath the roadway. Washington Roebling, who had just graduated from Rensselaer Polytechnic Institute in 1857, assisted his father in construction of the Allegheny Suspension Bridge, learning, firsthand, methods he would later use in building the Brooklyn Bridge.

21. Washington A. Roebling (1837–1926). At the age of 17, Washington Roebling was sent to the Rensselaer Polytechnic Institute in Troy, New York, where he was subjected to a strenuous course of study. Of the 65 students who started in his class, only 12 finished. His major interest, of course, was in the construction of suspended structures. This photograph was taken when he was 19 and still a student. **22. Washington A. Roebling, 1861.** Roebling joined the Union Army in 1861 as a private just as the Civil War was getting under way. During his tour of duty he served in the artillery and the cavalry, built four suspension bridges, and fought at Gettysburg, Bull Run, Chancellorsville and Antietam. He had risen to the rank of lieutenant colonel by the time he resigned his commission in December 1864, when he was 27 years old. His commanding officer and future brother-in-law, General G. K. Warren, commented, "Roebling . . . performed more able and brave service than anyone I knew." **23. Emily Warren Roebling (1844–1903).** In January 1865, Washington Roebling and Emily Warren were married. Emily was fated to play a major role in the construction of the Brooklyn Bridge after her husband suffered a serious attack of the bends while working beneath the New York Tower in 1872, an attack that was to render him an invalid for the rest of his life. Emily, seen here in a photograph of 1864, was able to grasp Washington's ideas and instructions well enough to interpret them to the assistant engi-

23

neers. Some days she would make two or three trips to the building site to check one or another of the many details involved in the construction of the bridge. An editorial of the day said Emily Roebling was the "chief engineer of the work," admired and respected by everyone who had anything to do with the bridge, including the assistant engineers, the bridge trustees and the local politicians.

24

24. The Cincinnati-Covington Bridge under construction, 1865. After Washington Roebling resigned from the army, he and Emily joined his father in Cincinnati to complete the Cincinnati-Covington Bridge spanning the Ohio River. This bridge had been started in 1856, but work was interrupted by the financial crisis of 1857 and the Civil War and was not resumed until 1863. In 1865, Washington took complete charge of the entire project. The central span of the bridge was 1,057′ and the width of the floor was 36′. The towers rested on timber platforms 17′ below the water where a bed of gravel and coarse sand afforded a good foundation. The two suspension cables were each 12½″ in diameter. This photograph shows the bridge under construction. The cables, already having been spun in place, are being wrapped with a tight wire binding. Roebling used this same method in building the Brooklyn Bridge.

25. The Cincinnati-Covington Bridge completed, 1867. This engraving from *Harper's Weekly* shows the Cincinnati-Covington Bridge shortly after it was opened in 1867. The accompanying article praised the new bridge while lamenting New York and Brooklyn's efforts to build their own: "The East River Bridge project has resolved itself into an impractical tunnel scheme, and for some time to come it's likely that the East River will be spanned only by bridges of ice."

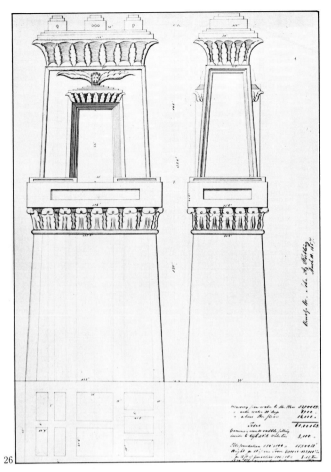

26

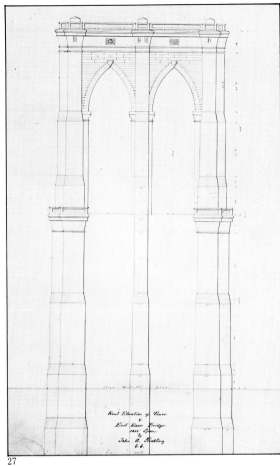

27

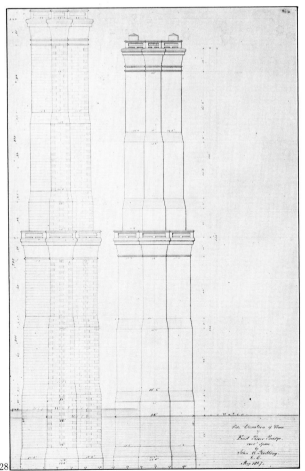

28

26. Early plans for the Brooklyn Bridge towers, 1857. During the winter of 1852, John Roebling was caught in an ice jam in the East River aboard a ferryboat. Washington, age 15, was with him and recounted that it was then that his father decided to promote the idea of a bridge spanning the East River. John Roebling made several drawings of the project including the one shown here: a massive Egyptian pylon with a winged lion's head surmounting the roadway entrance. Roebling signed this drawing in March 1857. **27 & 28. Preliminary plans for the Brooklyn Bridge towers, front and side elevations, 1867.** After considering a variety of architectural styles for the towers, Roebling decided on a Gothic double arch. These drawings for a 1,400′ span show the basic design with three different sets of dimensions. The final design of the tower was 8′ taller than the tallest one shown here and would accommodate a main span of 1,595′6″, almost 200′ longer than in Roebling's preliminary designs. **29. Design for Brooklyn Bridge tower, 1869.** An accurately detailed drawing indicates, block for block, how the tower would be constructed. The trusswork and cable-car tracks, which were to run across the center of the bridge, are also carefully rendered. The floor plan below shows that the tower would be three solid columns of granite joined below the roadway by hollow walls. After the towers were completed, an editorial in *Scientific American* suggested that these hollow areas be used as a distributing reservoir, since the Croton Reservoir at 42nd Street was unequal to the task of providing water for the whole city. The suggestion was not acted on, and in 1899, after the construction of the Croton Dam, the 42nd Street reservoir was demolished. It is now the site of the New York Public Library.

— Side Elevation — — Front Elevation —

M. H. Water

New York and Brooklyn Bridge
Masonry Tower
above M. H. Water

Scale 1 inch = 10 feet

— Plan at M.H.W. —

30

31

30. Bridge tour, Niagara Falls, April 14, 1869. To bolster support for the Brooklyn Bridge project, William Kingsley took a group of important Brooklyn citizens and influential engineers on a tour of the bridges built by John A. Roebling: the Monongahela, the Allegheny, the Cincinnati-Covington and the Niagara Falls railroad bridge. This photograph of the party was taken at Niagara Falls. The group stands on the suspension bridge designed by Samuel Keefer, opened just a few months earlier, which was the longest suspended structure in the world. It was, however, only 10′ wide, which made it rather impractical for carriage traffic. Some of the major participants here are: Julius Adams (left side of the picture), with a flowing beard, his elbow resting on the railing of the bridge; Washington Roebling, directly behind Adams, in military cap and full mustache. Directly behind Roebling is C. C. Martin, who became Washington

Roebling's chief assistant on the Brooklyn Bridge. Next to Adams is General Henry Slocum; and behind Slocum and slightly to his left is William Kingsley. Front and a bit right of center stands Horatio Allen with a large round face and top hat (after John Roebling's death, he would be appointed consulting Chief Engineer of the Brooklyn Bridge). Behind Allen and to the left stands B. H. Latrobe (with top hat and cane), Chief Engineer of the Baltimore and Ohio Railroad. To the right of Allen is William Jarvis McAlpine, president of the American Society of Civil Engineers. John Roebling stands on the far right wearing a light-colored topcoat. Three months later, John Roebling, the creative genius behind the Brooklyn Bridge project, would be dead. **31. Washington A. Roebling, ca. 1870.** In the summer of 1869, while John A. Roebling was engaged in fixing a location for the Brooklyn tower next to the Fulton Ferry Terminal, a ferryboat entering the slip pushed the pilings on which he stood in such a way as to catch and crush his foot. The injury resulted in lockjaw, and Roebling died 16 days later. Washington, who had worked closely with his father in the preparation of the design and plans of the Brooklyn Bridge, was appointed Chief Engineer. He was 32 years·old and would dedicate the next 14 years of his life to the bridge. **32. Cross section of the Brooklyn caisson, 1869.** In 1867, at the behest of his father, Roebling went to Europe especially to study the construction of pneumatic caissons, which can best be described as gigantic diving bells. It was on top of caissons that the towers were to be built. Workmen digging under the caissons would eventually sink them to a depth where there was a solid enough foundation to insure the stability of the towers. The Brooklyn caisson was made of timber and was 168′ long and 102′ wide. The inside chambers were 9′6″ high. Here the workmen would dig, sinking the caisson to a depth of 44′6″ below the bed of the East River. The roof was a solid 5′ of timber and the sides of the caisson tapered from 9′ at top to a narrow 8″ at the bottom. The bottom, called the "shoe," was well shod in iron. The excavated dirt, rocks and rubble would be shoveled into a pit below a square water shaft, then hauled to the surface by a clamshell scoop and deposited in a cart which would carry the debris to a waiting barge in the river. **33. Longitudinal section of the Brooklyn caisson, 1869.** The longitudinal section shows that the interior space was divided into six chambers. There were two water shafts (the largest openings), two supply shafts (the narrowest openings, only 2″ in diameter) and two air locks through which the workers passed to get in and out of the caisson.

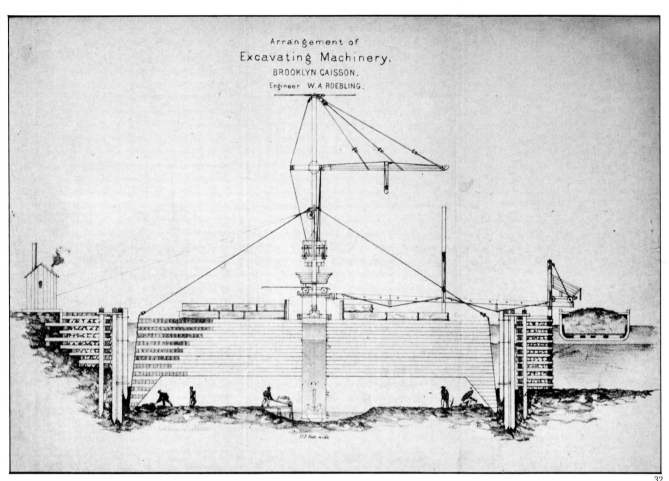

Arrangement of
Excavating Machinery.
BROOKLYN CAISSON.
Engineer W. A. ROEBLING.

102 feet wide

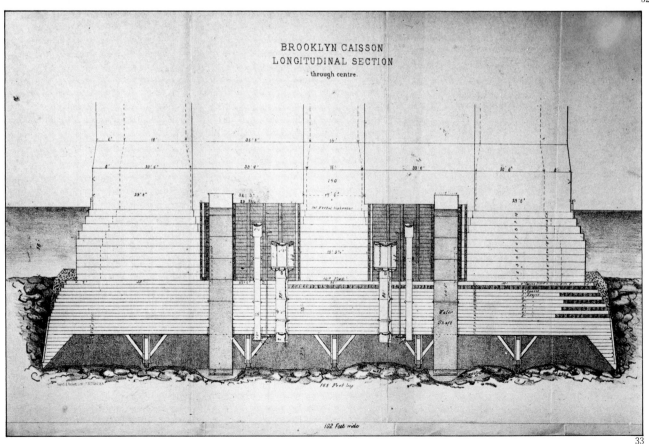

BROOKLYN CAISSON
LONGITUDINAL SECTION
through centre.

102 Feet wide

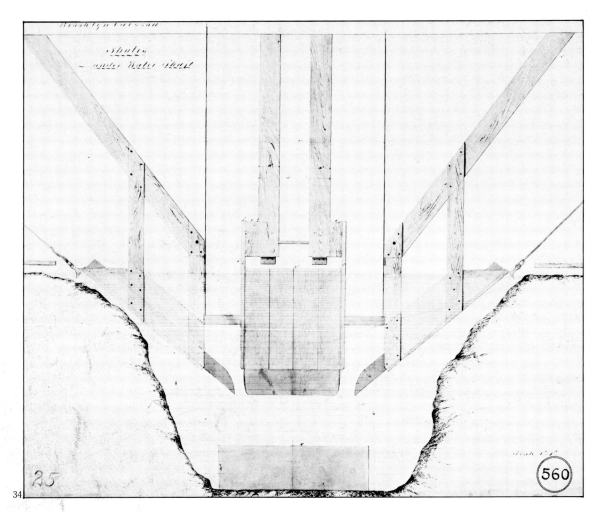

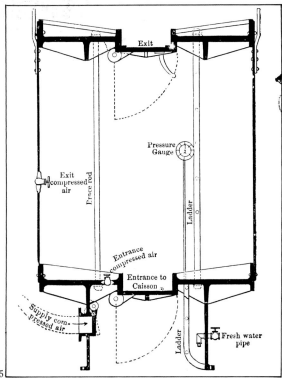

34. Arrangement of chutes surrounding the water shaft. This delicate, precise ink-and-watercolor drawing shows how wooden chutes were installed to guide excavated debris down under the water shaft where it could be picked up by a clamshell scoop (see

fig. 32). This drawing is one of over 11,000 engineering drawings and blueprints discovered in 1969 in a Department of Transportation carpentry shop located in Brooklyn, near the Williamsburg Bridge. These drawings document, step by step, the entire construction of the Brooklyn Bridge. Five hundred or more are signed by Washington Roebling himself. The complete collection is now part of the Municipal Archives of New York. **35. Airlock for the Brooklyn caisson.** The airlock through which the men passed to get in and out of the caisson was an airtight chamber with an iron hatch at the top and another at the bottom. In the Brooklyn caisson, men climbed down a ladder to reach the airlock. Once inside, the top hatch was locked and a valve allowing in compressed air was opened. When the pressure gauge indicated that the air pressure inside the lock and inside the caisson were equal, the bottom hatch dropped open and the men climbed down a ladder into the caisson. On leaving, the process was reversed. **36. Brooklyn caisson, exterior view, before launch.** The contract for building the gigantic caissons was given to Webb & Bell Shipyards at Greenpoint, Brooklyn. Work on the Brooklyn caisson started in October 1869 and was completed by the following spring. Three thousand people turned out to see the launching of this very strange-looking vessel on March 19, 1870. It could not be towed to its site, however, until a proper basin was dredged and surrounded on three sides by wooden piles. By May 3, this work was completed and the caisson was filled with compressed air and floated down the East River to its ultimate resting place beside the Fulton Ferry slip on the Brooklyn shore. Washington Roebling, William Kingsley, Horatio Allen and several others rode triumphantly on its deck. **37. Plans showing preparation for warping the caisson into its place.** The precise and delicate task of guiding the caisson into place is clearly depicted in this original engineering drawing signed W.A.R. by Washington Roebling, in September 1869. A simple but closely calculated system of cables, pulleys and winches slowly eased the caisson into its basin next to the Fulton Ferry.

36

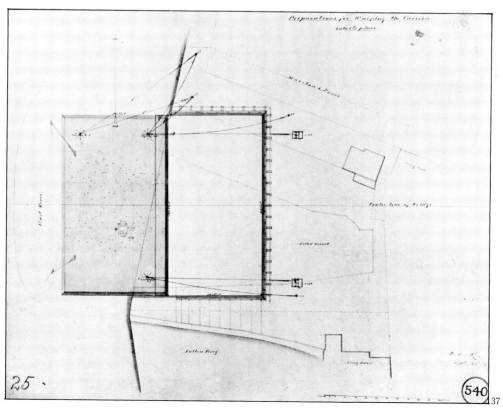

38

39

1.

2.

40

41

3.

4.

42

43

5.

6.

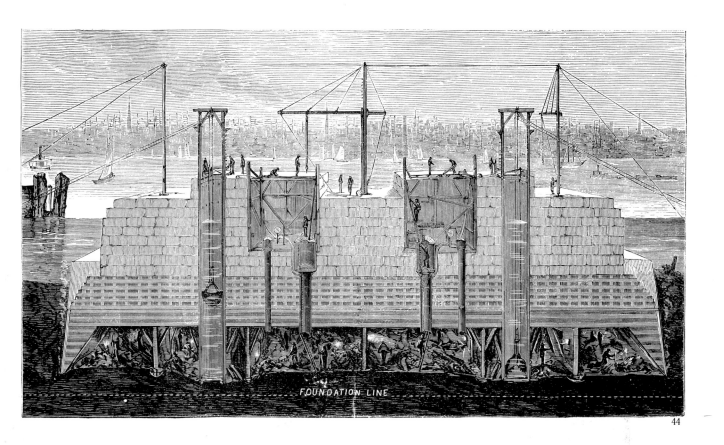

FOUNDATION LINE

44

38–43. Views of the interior of the Brooklyn caisson, 1870.
New Yorkers were fascinated by accounts of what went on inside the caisson below the surface of the East River. The illustrated newspapers obligingly printed wood engravings, made from their artists' sketches, of the caisson interior. The first picture in this series shows the air lock where three gentlemen, of the press perhaps, wait for the air pressure to be equalized before climbing through the bottom hatchway and eventually into the caisson (panel no. 2). The interior of the caisson was described by one of the workers, Master Mechanic E. F. Farrington, in this way: "Inside the caisson everything wore an unreal, weird appearance. There was a confused sensation in the head. . . . The pulse was at first accelerated, then sometimes fell below the normal rate. The voice sounded faint unnatural, and it became a great effort to speak. What with the flaming lights, the deep shadows, the confusing noise of hammers, drills, and chains, the half-naked forms flitting about, with here and there a Sisyphus rolling his stone, one might, if of a poetic temperament, get a realizing sense of Dante's inferno." The interior was lighted by 14 limelights which were fed by gas piped from the outside. The temperature was always hot, at least 80° and extremely humid. No. 3 shows workmen breaking up a rock. Rocks and boulders, some 14′ long and 5′ wide, became such a problem in sinking the caisson that eventually explosives had to be used. Workmen would drill holes in a rock into which a charge would be placed and then set off. Men would then break up the larger fragments (no. 4) and haul them off to the water shafts shown in no. 5. Rocks, debris and other

excavated material would be dumped into the pool beneath the water shaft where it was lifted to the surface by a clamshell scoop. Panel 6 shows workmen sawing timber wedges which were placed beneath the caisson's partitions. As several inches of earth were cleared from beneath the shoe, these wedges were knocked out of place from beneath the partitions, thereby lowering the caisson gradually and evenly. **44. Cross section of the Brooklyn caisson, 1870.** A feature published two months later included this cross section of the entire caisson showing workmen climbing in and out of the air locks, the men above working the derricks which have already laid eight courses of stone for the towers. One can see how tightly the riverbed fits around the caisson, sealing off all water. The level of water in the water shafts was gauged to seal in the compressed air effectively without flooding the interior of the caisson. On one occasion, when the water level had become too low and consequently was not heavy enough to contain the air pressure inside, there was a tremendous blowout, that is, compressed air rushed up the water shafts, carrying with it water, rocks, mud and other excavated debris and hurling it 500′ into the air. Because it was Sunday, there was no one working in the caisson and only three men on duty above, one of whom was hit by a rock. The shafts were quickly refilled with water and compressed air pumped back into the caisson, but for the duration of the mishap, the weight born by the caisson's outside walls and inside partitions, was 23 tons per square inch—yet there was no damage. Roebling had designed the caisson to hold 5 tons per square inch.

45

45. Exterior view of the Brooklyn caisson, 1870. In December 1870, a fire that started inside the Brooklyn caisson threatened the future of the entire undertaking. A workmen had carelessly held his candle near the highly inflammable oakum caulking between the beams of the ceiling. The flame burned up and into the wooden structure. Because of the oxygen-charged compressed air, it burned so rapidly that it defied all ordinary fire-fighting methods. Roebling finally ordered that the caisson be flooded to make absolutely sure that the fire was extinguished. The fire caused a three-month delay, since all the damaged wood had to be cut out and replaced. But by March 6, 1871, with the restoration work complete, Roebling ordered that the caisson, at a depth of 44′6″, be filled with concrete, the last step in the creation of a solid foundation for the tower. It was during this arduous effort to extinguish the fire that Roebling suffered his first attack of caisson disease or the "bends." After spending an entire night working inside the caisson, Roebling collapsed and had to be carried to his home in Brooklyn Heights. By the next day, however, he had recovered.

46. **The New York caisson.** The New York caisson, also built at the Webb & Bell Shipyards, was launched on May 8, 1871 and was towed into its permanent home in September. It was basically the same as the Brooklyn caisson with some changes to accommodate what had been learned from the first experience. It was, for instance, lined with boiler plate to prevent fires.

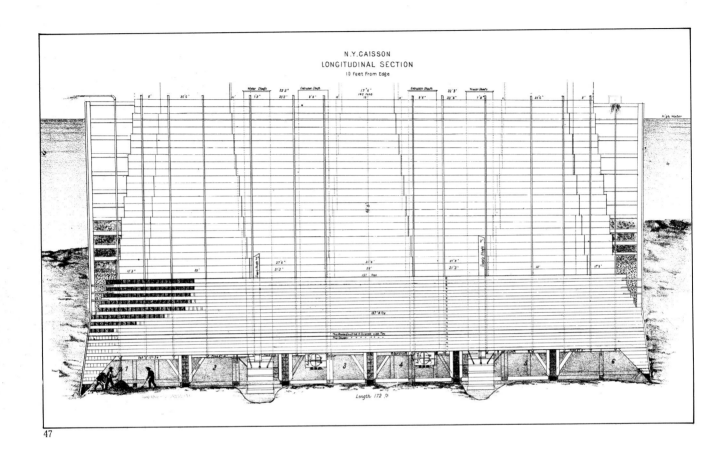

N.Y. CAISSON
LONGITUDINAL SECTION
10 Feet From Edge

47. Longitudinal section of the New York caisson. By means of compressed air, sand and gravel could be drawn up narrow pipes through the roof of the caisson. Roebling had about 50 of these pipes installed, greatly facilitating the task of lowering the caisson. This drawing, which was part of Roebling's report *Pneumatic Tower Foundations of the East River Suspension Bridge*, shows workmen in the lower left corner feeding a sandpipe. At the top of the pipe a metal elbow directed the sand and gravel toward a receiving scow. The force of the sand traveling through the pipe quickly ate right through the metal elbow and the engineers had to use instead a block of granite which they positioned above the pipe. Roebling's report mentioned that a boatman on the river had his finger shot off by a misguided piece of gravel and a bridge laborer was shot right through the arm by a large fragment.

48

48. New York caisson, March 1872. By March 1872, the New York caisson rested at a depth of 68′. Its relatively easy descent through quicksand was suddenly slowed as it approached bedrock. In April, at a depth of 71′ the first death that could be attributed to the bends occurred. By May there were several more deaths and Roebling himself had collapsed a second time and had to be carried home. As Emily Roebling noted in her diary, it was a time of intense anxiety for him. Finally, on May 18, 1872, he called a halt to any further excavation. The caisson rested at a depth of 78′6″ and he felt sure that, although they had not reached bedrock, they had reached a stratum of sand and gravel "good enough to found upon, or at any rate nearly as good as any concrete that could be put in place of it," as he wrote in his report to the trustees. In this photograph barges laden with granite blocks rest off to the side. In the center, across the river, is the Brooklyn tower now at a height of about 80′. To the right of the tower is the newly completed Fulton Ferry House.

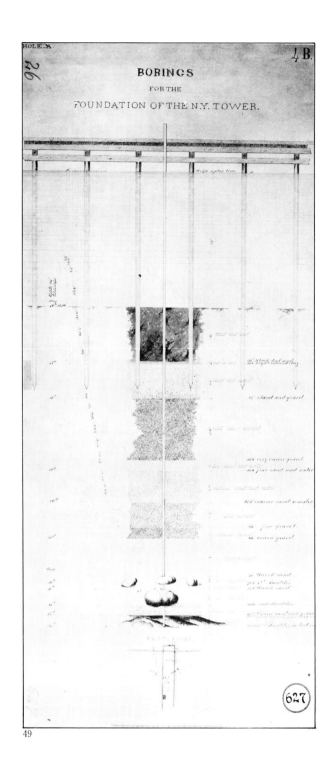

49. Borings for the foundation of the New York tower. This watercolor, recording borings made from May 16 to June 25, 1870, shows, layer by layer, the composition of the earth beneath the New York tower. The top layer is made up of black dock mud, followed by clay, alternating layers of sand and gravel, quicksand, boulders, and finally bedrock at 84′4″ below the high-water line of the East River. The New York tower rests 78′6″ below high water, 6′ above bedrock.

50. Bird's-eye view of New York, 1870. This view of Brooklyn and New York joined by the magnificent East River Bridge was published before the bridge was built. In November 1870, the only work under way was well below the surface of the East River, inside the Brooklyn caisson. This view gave New Yorkers an exciting vision of how the city would look when the bridge was completed. Along with this futuristic view *Harper's* included a word of caution: "The only doubt seriously felt [concerning the bridge]—though not shared at all by the Roeblings, father or son—is in the stiffness of the suspended structure against the lifting, twisting and swaying effect of powerful winds." The lower right-hand corner shows Printing House Square. The tall spire on the right is Trinity Church. On the Brooklyn side the spire of Holy Trinity Church marks the corner of Clinton and Montague Streets. Just beyond that is City (now Borough) Hall, and the County Courthouse.

Brooklyn Tower
Sept 1872

51. Brooklyn tower, September 1872. In the fall of 1872, S. A. Holmes, a noted New York photographer, was commissioned to record the progress of the bridge towers. The Brooklyn tower had reached the level of the roadway—a height of 119′. The engineering crew and workmen pose proudly amid a forest of scaffolding and boom derricks. The vertical tracks running up the sides of the pier guided the blocks of granite as they were hoisted to the top by huge iron pulleys which were powered by steam engines located to the rear of the structure. **52. Brooklyn tower, September 1872.** The photographer, S. A. Holmes, identifies the man standing on the left looking out over the river as E. F. Farrington. However, he has also been identified as C. C. Martin, Roebling's assistant, or as Roebling himself. By this time, Roebling was quite ill. After his last attack of the bends, he had lain near death for days. Although he returned to work for a time, he never recovered his full vigor and relapses occurred frequently until, by December 1872, he had to take a leave of absence. He was not to set foot on the bridge again until after its completion. Others identified are: (2) William C. Kingsley, who had been appointed General Superintendent, a title that would cause him some embarrassment later on; (3) O. P. Quintard, bookkeeper; (4) "Alex" McKinnon, foreman of the masons on this pier; (5) F. Collingwood, assistant engineer and friend of Roebling's from R.P.I.; (6) George McNulty, assistant to C. C. Martin; (7) T. G. Douglas, head mason, who one year earlier, in October 1871, had been in a terrible accident on the Brooklyn tower when one of the guy wires supporting a boom derrick broke, hurling two derricks down from the tower. Douglas was pinned under one of the derricks and suffered great damage to his kidneys. Although he continued working part time after the incident, he never regained his full health, and died of his injuries two years later. (8; below, hidden in the scaffolding) W. H. Paine, assistant engineer in charge of the New York tower.

53

53. New York tower, September 19, 1872. Across the river, to all appearances, the New York tower was just getting under way, although, as Roebling pointed out in his report to the trustees, the masonry in the foundation below the surface of the river equaled the entire visible part of the Brooklyn tower. Despite the massive solidity of the two towers in progress, the press still had serious doubts about the stability of the bridge. The *New York Times* reported, "Some think that a third pier would have to be built in the middle of the river." As spokesman for the bridge committee, William C. Kingsley assured the *Times* that the bridge would be absolutely stable and would not be swayed even by winds of 60 mph.

54. New York tower, September 19, 1872. A long shot of the New York tower shows the worksheds housing the steam engines which powered the hoisting equipment on top of the tower. Notice the coal waiting to be hauled up the track. On top of the tower are three granite blocks being maneuvered into position by the giant boom-derricks. A glimpse of the Brooklyn pier can be seen over the top.

55

57

55. Brooklyn, October 1872. This view from the top of the Brooklyn tower shows where the bridge's approach would eventually be built. All the buildings on the center line of the photograph were demolished, including St. Ann's church, which stood at the corner of Washington and Prospect Streets. St. Ann's was torn down in 1880. The Union Building, the light-colored structure just right of center, was home for the Brooklyn *Union* newspaper; it also housed the offices of the Bridge Company. **56. "How Kingsley and Fowler Amused Themselves in Spending the Money for Bridging the East River."** The Bridge Company was constantly being assailed by the press in regard to the management of its financial affairs, especially during 1873, when the notorious Tweed Ring was broken and certain allegations came to light concerning Tweed's position as member of the Executive Committee of the New York Bridge Company and a major stockholder. Kingsley was also a major stockholder and General Superintendent of the bridge. For his services he was being paid 15 percent of the amount of the construction expenses, meaning that during the year of 1870 alone he was paid $175,000. According to the Bridge Company's minutes, this was done at Tweed's suggestion. By 1873, in a new atmosphere of reform, the Board of Trustees of the Bridge Company renegotiated Kingsley's salary to a flat $10,000 a year. William Fowler, who is the New York pier in this political cartoon, had nothing to do with the Brooklyn Bridge. Fowler was one of the Commissioners of the Brooklyn Board of City Works and was involved in another scandal with Kingsley regarding his role in the building of the Hempstead Reservoir, for which Kingsley & Keeney were the contractors. **57. William Marcy Tweed (1823–1878).**

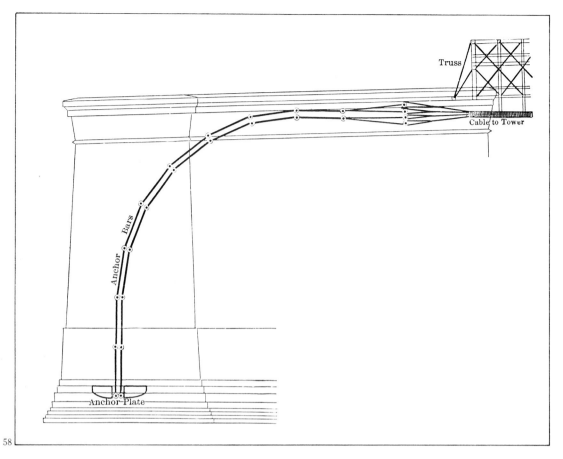

Truss

Cable to Tower

Anchor Bars

Anchor-Plate

58

59

60

58. Section of the top and back of anchorage—side view.
Immense anchorages on either side of the bridge hold the great cables in place. Each measures 90′ × 119′ × 132′, weighs 60,000 tons and is made of granite and limestone blocks. Beneath the structure are buried four massive anchor plates of cast iron, one for each cable. This diagram shows a side view of an anchor plate in place beneath the anchorages and its double chain of eyebars which would eventually hold one of the main cables. **59. Casting the anchor.** Work began on the Brooklyn anchorage in February 1873, and by May the first of the anchor plates was successfully cast. With each anchor measuring 17½′ × 16′ × 2½′ and weighing over 46,000 pounds, just transporting and placing them in the right position was a major logistical feat. **60. The Brooklyn anchorage, 1873.** Each anchor had two rows of nine oblong openings into which 18 wrought-iron eyebars would be placed. A large steel pin which slipped through the "eyes" of the eyebars below the anchor holds them in place. By summer, all four plates in the Brooklyn anchorage had been installed and the work of constructing the double chain of eyebars and building up the masonry around them could begin. It would take over two years to complete the job. In the background of this engraving is the Brooklyn tower, its steam-powered boom-derricks hoisting granite blocks into place.

61

61. View from the Brooklyn tower, 1873. The Brooklyn tower, completed to above the level of the roadway, was now climbing skyward in the form of three separate piers which eventually would be joined at the top to form two huge Gothic arches. *Harper's Weekly* surmised that few readers would care to climb to the top of the tower for this "magnificent panorama" of New York harbor, and therefore sent its artist to record the scene. In the distance are the spires of Trinity Church and St. Paul's Chapel. Toward the foreground is James Bogardus' shot tower, and to the right of it is the new Post Office under construction. This building stood at the southern tip of City Hall Park until its demolition in 1938. **62. New York tower, September 1873.** 1873 had been a difficult year for the Bridge Com-

pany: not only had it been assailed by scandals and investigation but the City of New York was refusing to pay its share for the construction of the bridge. As Mayor Havemeyer of New York ingenuously explained: "There was some reason for the Cincinnati-Covington Bridge because it opened up immense tracts of land. That however is not the case with the Brooklyn Bridge because they could not travel more than 5 miles without hitting the ocean." It seemed to him that one of the objects of the Brooklyn Bridge was to make the lots around Prospect Park more valuable. Despite all the problems, work on the New York tower had progressed steadily so that by September it was completed up to the level of the roadway.

63

63. The Brooklyn tower, 1874. Work progressed steadily, although not rapidly. By 1874, when the Brooklyn tower was rising like a giant granite pitchfork above the four- and five-story houses of Brooklyn, Roebling, who had spent six months in Wiesbaden, Germany, returned to his home in Trenton, New Jersey. From there he drew up all specifications for building materials and sent precise, detailed instructions to his engineering staff regarding the work at hand. But he was still very weak and suffered a great deal of pain. He did not return to New York until 1876. **64. Brooklyn tower, 1875.** Further along in its gradual evolution, the Brooklyn tower begins to assume its final shape, an imposing Gothic double arch. The wooden constructions within each arch are falsework, supporting the masonry until it was all laid. The tower was completed in June 1875. **65. The U.S.S. *Swatara* passes the piers, October 30, 1875.** This wood engraving shows the Brooklyn tower

complete and the New York tower just short of completion; it was finished in July of the following year. The *Swatara* was on her way to Pará, Brazil, to rescue former Confederate soldiers who had fled the United States after the Civil War. The United States government, upon learning that the former soldiers had fallen on hard times, sent the ship to bring them home. The *Swatara*, whose tall masts would not have fitted under the completed bridge, was one of the ships cited by the New York Council of Reform as an example of how the bridge would obstruct maritime traffic on the East River. The Council maintained that it would be folly to spend more money "on a bridge that is not called for . . . very seriously damages a large part of commerce of this harbor, taxes the financial ability of these two cities to their utmost and cannot fail either to be taken down by the mandate of the courts or demolished by the wind."

64

65

45

66. The Beal Panorama of Manhattan, 1876. By 1876 both towers were completed. This sweeping panorama of Manhattan, taken from the Brooklyn tower by Joshua Beal, is actually five separate images joined together to make a composite view. New Yorkers had never seen their city like this before. All the major points of interest are visible (Trinity Church, the Western Union Building, which was completed in 1875, St. Paul's Chapel, the shot tower, the new Post Office and the new Tribune Building, also completed in 1875). Dominating the whole scene is the stark and massive New York tower. New York architects shuddered at the thought of the bridge being left incomplete (with all the controversy surrounding the bridge, completion was often in doubt) and its towers remaining in "unnecessary ugliness." **67. The New York anchorage, July 1876.** This photograph shows the New York anchorage in the foreground with the New York tower and the Brooklyn tower in the background. Work on the New York anchorage started in May 1875. The spot on which the anchorage rests marks the original shoreline of Manhattan. When excavating the site, workmen found many artifacts dating from the 1600s—coins, rings, cannonballs and a bayonet. Farther down were a number of freshwater springs, necessitating the installation of a powerful pump that was kept running day and night to prevent the foundation from being flooded.

66

67

Dated *New York 14* 187*6*

Received at *17.45 August*

To *Col W. A. Roebling*

West State St.

The first wire rope reached its position at Eleven and one half oclock. Was raised in Six Minutes.

W H P

18 Paid Am

69

70

68. New York anchorage, July 31, 1876. All is in readiness for the first wire rope to be stretched across the East River. On August 11, after waiting for traffic on the river to subside, a ¾″ steel-wire rope that had been unwound and dropped to the bottom of the river was pulled taut between the two towers. The great event was watched by a large crowd of sightseers and was reported at great length. It was the first tenuous thread to join the two cities after seven years of preparing the towers and anchorages. **69. Telegram to Washington Roebling.** The triumphant achievement was telegraphed to Roebling in Trenton by W. H. Paine. That same afternoon a second rope was also raised into place. At each anchorage the ends of the ropes were spliced together around 12′ drums to form an endless rope, which would be used to hoist more and heavier cables into place. **70. Farrington crosses the East River, August 25, 1876.** The next dramatic occasion followed shortly thereafter. An endless traveler rope had been installed that could be driven by a steam engine located on the Brooklyn anchorage. Master Mechanic E. F. Farrington was chosen to be the first man to ride the traveler rope across the river. Seated on a boatswain's chair as he started out from the Brooklyn anchorage, Farrington was greeted by the shouts and cheers of an immense crowd that had come out to witness the

historic event. When he was over the river, ferryboats, tugboats and steamers all blew their whistles to cheer him along. Oblivious to danger and elated by the experience, Farrington doffed his hat to his well-wishers below. Twenty-two minutes after he left the Brooklyn anchorage he arrived safely atop the New York anchorage. **71 (Overleaf). The footbridge.** By February, five temporary wire cables had been lifted into place, two supporting a footbridge to be used by the workmen in laying the permanent cables. The construction of the footbridge was supervised by E. F. Farrington, a man well familiar with the work from his experience on the Cincinnati-Covington Bridge. Five cradles were also hoisted into position. They were 100′ long and 5′ wide. The operation of building the footbridge and installing the cradles was a hazardous one, watched by great crowds on both sides of the river. As one New York paper observed, "Much admiration was expressed at the skill and daring of the workmen." Washington Roebling watched too, through a high-powered telescope mounted in the window of his home on Columbia Heights. Roebling had returned from Trenton in the fall of 1876 to see, for the first time, the completed towers that held aloft the first wire ropes strung between them. Roebling said it looked just as he imagined it would.

72

72. April 1877. This view from the Brooklyn side shows the footbridge and four of the five cradles. These structures look exceedingly slender and unstable, especially when contrasted with the massive towers. Roebling himself remarked in his report to the trustees that in order to "protect such a frail structure [the footbridge] against the violence of terrific gales that rage here almost weekly, is no easy task, and I venture to predict that notwithstanding every precaution, our temporary walk will be disabled more than once before we have completed cablemaking." **73. Top of the Brooklyn tower and footbridge, 1877.** For a time, the footpath was open to tourists, and all one had to do to gain access was ask for a ticket at the bridge office. Eventually the number of sightseers became so numerous that this privilege was discontinued. Workmen often had to be called away from their jobs to rescue a terrified man or woman immobilized by fear of the dizzying height and unsettling sway of the footbridge. Women, however,

seemed less affected by the experience than men. As one workman was quoted in the *New York Times*, "The women are soonest at ease, and you'll see them swinging their parasols carelessly where brave men hold on with both hands." There was also some vandalism; people actually cut the cables to carry off strands of wire as souvenirs. **74. Top of the Brooklyn tower and footbridge, 1877.** Another view of the footpath was printed in *Harper's* rival newspaper, *Frank Leslie's Illustrated News*, during the summer when the bridge was at its peak of popularity. Henry Murphy, President of the Bridge Company, personally issued the passes and decided who could go and who could not go. He declined to issue a pass to a budding young actress who wanted to ride across on horseback, but he did allow a married couple to carry their newborn infant across; Murphy himself never ventured onto the footbridge. As he explained to a reporter, he decided that the Bridge Company could not yet afford to lose its president.

73

74

75

75. Brooklyn anchorage, 1877. This official-looking party, perhaps trustees of the Bridge Company, are posed just beyond the solid safety of the Brooklyn anchorage. Note the warning prominently posted for visitors starting their journey over the footbridge. By the middle of August several thousand people had crossed, including four-year-old Al Smith, whose father worked for the Bridge Company as a guard. By September of 1877, after a man suffered an epileptic fit on the footbridge and had to be tied down until his seizure passed, the issuance of passes was finally stopped entirely.
76. Spinning the cables, 1877. On June 11, 1877, the process of spinning the four main cables was ready to begin. Each cable would have 19 strands of 278 wires each. The wires were not twisted but laid up parallel. The strands were formed precisely like skeins of yarn. A carrier wheel (see Figure 75; upper left) was attached to the same traveling rope which had conveyed E. F. Farrington across the

river the previous summer. A loop of wire placed around the wheel was carried across both towers to the New York anchorage, where workmen would grab the loop and slip it around a heavy iron shoe at the rear of the anchorage. The empty wheel would then be sent back to the Brooklyn anchorage to pick up another loop. Thus each trip made by the carrier wheel would lay two wires. Men positioned in the cradles and on the towers made sure that the length of each strand was properly regulated, no easy task as the wires lengthened or contracted during the day according to fluctuations in temperature. **77. Binding the strands.** After enough wires to make a strand were laid out, workmen would go out in hanging platforms called "wrapping buggies" to bind the strands with wire at intervals of 15″ from anchorage to anchorage. **78. Saddle plates.** As the wires passed over the tower, they were temporarily held 3′ above the saddle plates on rollers. After all the wires were properly regulated

76

77

78

and bound into a strand, a crew of 10 men operating a capstan would lower the strand onto the saddle plates. **79. Section of tower, showing saddle plates and lowering of strand into position.** Additional rollers were located beneath the saddle plate. These made the saddle movable to allow for variations of stress on the cables so they wouldn't slip, chaff or exert any lateral strain on the towers. In fact, however, the rollers never worked. Because of the tremendous weight of the cables, the rollers soon became embedded in the iron plate on which they rested. However, the towers were saved from damaging lateral strain by the wooden caissons which provided their foundation. The thickness of the wood afforded enough of a cushion for the towers to rock on their bases, thus alleviating strains on the masonry from unbalanced cable-pull.

79

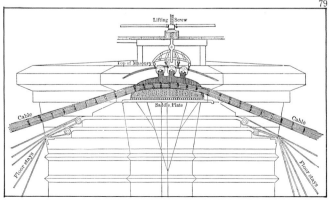

80

80. Side view of cables, Brooklyn anchorage, 1877. The strands had to be lowered into place at the anchorage. The heavy iron shoes that held the strands were temporarily located at the back of the anchorage throughout the spinning process. As soon as a strand was completed, its shoe would be lifted by a powerful block and tackle to the front of the anchorage, thus lowering the strand into place where it would eventually be wrapped together with the other 18 strands to form one of the four main cables. A 7″-long steel pin fastened the shoe to the eyebars which projected out of the anchorage. Thus, each strand was permanently fastened by a chain of eyebars to the 23-ton anchor plate located beneath the anchorage. **81. Elevation of Brooklyn anchorage. 82. Top view of anchorage.** Inside a huge shed built atop the Brooklyn anchorage were 32 drums, each measuring about 9′ in diameter and 2′ wide, and holding 10 miles of wire which had been previously oiled and spliced in the yard behind the Brooklyn anchorage. Every drum was equipped with spokes on one side so that the paying out of wire could be regulated by a team of workmen. Two large grooved wheels, powered by steam engine, were located at the front of the anchorage. Each drove a traveler rope and carrier wheel which, in turn, pulled a loop of cable wire from anchorage to anchorage. **83. Arrangement of cable drums, footbridge, cradles and cables on the Brooklyn anchorage.** This illustration gives a very good idea of how the whole system worked. It shows the drums (two sets of four for each cable); the steam-powered grooved wheels (*a, b, c, d, e*), which drove the traveler rope (*A*), the cables which supported the cradles and footbridge, and two cables in the process of being drawn across the bridge by carrier wheel (*G*). When the traveler ropes passed over the towers, they were drawn over sets of grooved cast-iron wheels set upright and revolving between large timbers bolted together and extending across the towers. Likewise, when the ropes passed over the cradles, they were run on top of small grooved wheels set in a frame. The whole system was driven by a 20-horse-power steam engine stationed at the foot of the anchorage.

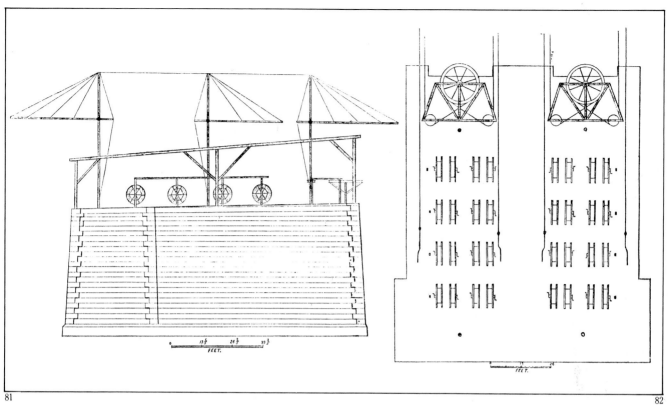

81 82

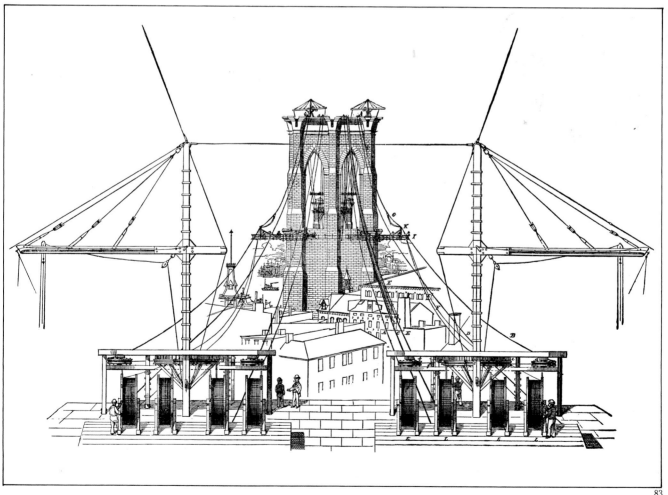

83

85

84. Brooklyn anchorage, 1878. A view from the Brooklyn tower to the anchorage shows a number of completed strands. On top of the anchorage, obscured by a protective tarpaulin, is the system of wheels which guided the traveler rope, and the shed, which housed 32 drums of steel wire. St. Ann's, slated for demolition in 1880, still stands in the distance, and a blur of streetcars rushes along Fulton Street. **85. Brooklyn anchorage and tower, 1878.** This view, looking from the anchorage up to the tower, shows the massive clusters of eyebars holding the first laid strands. On the right is a finished strand ready to be lowered into position, and next to the footbridge is a new sign: "No Admittance Here." **86. Accident on**

the New York anchorage, 1878. On June 14, E. F. Farrington was supervising the lowering of a strand when the wire rope holding the shoe suddenly snapped, catapulting the shoe and the strand attached to it into the yard behind the New York tower. The weight of the unrestrained strand then yanked the shoe up and over the tower where it finally crashed into the middle of the East River, narrowly missing a crowded ferryboat. Two of the four men working with Farrington were killed and the other two were seriously injured. Farrington was unhurt except for a scratch on his hand. It was the most terrifying accident to occur during the construction of the bridge.

84

86

87

87–91. Making wire for the bridge, 1877. J. Lloyd Haigh of Brooklyn was awarded the contract for manufacturing the steel cable wire. These wood engravings accompanied an article in *Scientific American* describing in detail the precise and closely calculated procedures used in making the wire, which was supposed to have a breaking strength of 3,400 pounds. All the wire had to be straight wire. That is, when it was unwound, it had to lie flat with no tendency to spring back. It had to be thoroughly galvanized in a bath of muriatic acid heavily charged with zinc to protect it from the elements. And before it was sent to the Brooklyn anchorage, the wire had to be carefully tested under the supervision of the bridge engineers. But despite all of these precautions, in July 1878, Roebling informed Henry Murphy that, after some investigation, his engineers had discovered that Haigh had been smuggling substandard wire into the bridge and that all of it was now irrevocably woven into strands along with good wire. Roebling had no idea how much bad wire had been used, but since the cables had been designed to be six times stronger than needed, it was decided that the bridge would still be safe regardless of the bad wire. Haigh retained his contract and under close surveillance provided enough extra good wire to add 150 more wires to each cable.

MAKING the LARGE COILS

88

TESTING the WIRE

89

SIZE OF THE WIRE

90

GALVANIZING the WIRE

Bross

91

92

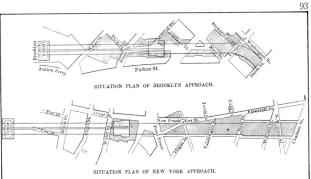

SITUATION PLAN OF BROOKLYN APPROACH.

SITUATION PLAN OF NEW YORK APPROACH.

93

92. Brooklyn anchorage, October 1878. The man in the center wearing a top hat and with arms crossed is Henry Murphy. If he looks grim, it is easy to understand why. New York City was refusing to meet its obligation of half a million dollars, slowing work on the bridge considerably. Also, a rumor had been circulated that a stonemason had secreted some dynamite in one of the towers and would blow up the works sometime soon. Finally, there was the wire fraud. In September, the contract for wrapping wire was quietly withdrawn from J. Lloyd Haigh and awarded to John A. Roebling & Sons.

However, a more pleasant occasion is commemorated in this photograph: the completion of laying the strands. The last wire went across on October 15, 1878. The figures, starting second from the left are: John H. Prentice, treasurer; O. P. Quintard, secretary; Henry C. Murphy; Van der Bosch, draftsman; William H. Paine; George McNulty (seated at right, wearing a top hat). The other five men identified, standing in front of the drum shed, are from the right: Wilhelm Hildenbrand, chief draftsman; Francis Collingwood; C. C. Martin; E. F. Farrington; and William Dempsey, foreman of the riggers. **93. Plans for the approaches.** The shaded areas on these street maps indicate the property that had to be cleared to build the approaches to the bridge. The cost of this land was $3,800,000. The New York approach, one-third longer than the Brooklyn approach, starts at Park Row just opposite City Hall. The Brooklyn approach curves up from Sands Street. **94. Demolition for the New York approach, 1877.** This engraving, which appeared just as demolition was getting under way, shows the New York anchorage with two bridge towers in the background. Workers are removing debris from the buildings which stood between Franklin Square, Frankfort and Cliff Streets. The anchorage itself sits on the site of George Washington's first executive mansion at 1 Cherry Street. Washington lived there from April 1789, after his inauguration at Federal Hall on Wall Street, until February 1790, when he moved to a larger residence on lower Broadway.

95

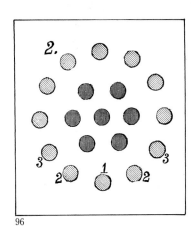

2.

3

3

2

1

2

96

3.

2

2

97

4.

A

A

98

99

95. Cable making, 1878. As the wires were bound together into strands and lowered into place, the strands were, in turn, bound together to form manageable bundles. Eventually, all 19 strands would be joined, but the first step was to bind up what was called the core—the seven middle strands. **96–98. Diagram of cable making, 1878.** No. 2 shows the position of the 19 strands (which were laid up bottom ones first—not starting at the center, which the circular arrangement would seem to imply). It also shows the central seven strands, which were the first to be bound together. At this point, 12 strands had been laid, allowing the workmen to start the process of binding the core. First the strands were hauled up close vertically (no. 3) by means of a rope and derrick tackle and then held with a wooden clamp every three feet. After wires which

initially bound the strands had been removed, a massive iron clamp (no. 4) compressed all the strands into an even cylinder. In Figure 85 you can see one of the iron clamps resting on the floor of the anchorage next to the eyebars. **99. Cable wrapping, 1878.** After the core was bound, the upper strands, as fast as they were completed, were lowered into place. The core lashings were taken off one by one, and the whole cable was bound together in precisely the way described previously, only using larger clamps. In this photograph, looking up from the Brooklyn anchorage to the Brooklyn tower, we can see the larger clamps in place. In November 1878, Henry Murphy had to shut down all work on the bridge because New York City refused to pay its share of the bridge costs, and wrapping was not resumed until March 1879.

100

100 & 101. Wrapping the cables, 1878. After the 19 strands were assembled into cables, they were bound with a close spiral wrapping of wire. This was done by a machine which consisted of an iron drum bolted around the cable. The wire, held in severe tension, wound off the drum and around the cable to form a tight wrapping. One machine could wrap 10′ of cable in a day, and 16 cable-wrapping machines were used simultaneously, four per cable. *Scientific American* observed: "These operations, though simple in themselves, acquire a special interest from the circumstance that they are carried on at such a gigantic scale and such an enormous elevation above the river." **102. Cable wrapping.** Another view gives some idea of what it must have been like to work during inclement weather, with high winds buffeting the workmen's suspended platform. New Yorkers watched this activity with great curiosity and admiration. The workers themselves seemed quite nonchalant about the dizzying heights at which they worked and the danger of high-velocity winds.

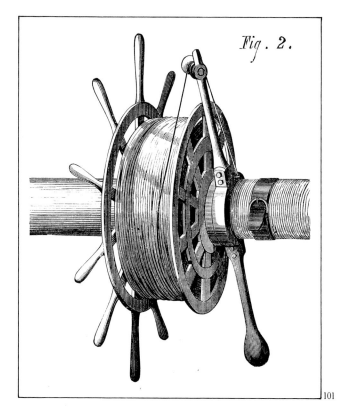

Fig. 2.

101

102

103

104

103. New York tower, 1879. In March 1879, the courts ordered the city of New York to pay its share of the financial obligation for the bridge. Construction work, which had been stopped for six months, was immediately resumed, and 605 men were rehired to finish the cable wrapping and the masonry on the approaches. That spring, another suit brought by a group of shipowners to have the bridge torn down as an obstruction to shipping along the East River was also settled, once again in favor of the Bridge Company. Roebling, however, was under fire from the Board of Trustees, with one member, Gen. Henry Slocum, charging that the bridge engineers were taking bribes from steel manufacturers. This charge was prompted by Roebling's decision to use steel instead of iron for the trusswork

of the bridge floor. An investigative committee was appointed, and for the next month rancorous accusations were printed in the daily newspapers. Roebling was furious—and rightly so. As it turned out, there was no basis for Slocum's charges whatsoever. The cause of the entire misunderstanding was a visit Col. W. H. Paine had paid to a steel company in Pittsburgh to observe a new method of making steel eyebars. Other steel companies misconstrued the purpose of Paine's visit. Their speculations became rumors, and the rumors gave rise to the trustee's unwarranted charges. **104. Brooklyn tower, 1879.** In this view from the Brooklyn anchorage four of the suspended platforms used for wrapping cable can be seen. One can even see the ladder leading from the platform to the cradle.

105

105. New York anchorage, 1879. The eyebars and cables are now completely encased in 60,000 tons of stone. In his report three years earlier, Roebling pointed out that in European bridges the anchor chains were usually carried in tunnels open to access, their care being entrusted to future generations. "This, however, is contrary to the genius of the American people," he wrote, "with whom everything has to look out for itself; hence, in the arrangement of this anchorage, the chains are inaccessibly preserved and are not entrusted to the neglect of posterity." **106. South Street and the New**

York anchorage, 1879. A long shot of the anchorage includes a detailed view of South Street with its shops and ships in a quiet break from its usual frenzied mercantile activity. In the background is the Post Office, which stood in City Hall Park from 1875 to 1938. To the right of the Post Office is the New York Tribune Building (1875), which stood two blocks south of the bridge entrance until it was demolished in 1966. The long, low building visible below the Tribune Building housed the offices of Harper's.

108

107. **Drawings of Collingwood, 1877.** Although Wilhelm Hildenbrand is usually credited with designing the approaches, these drawings, which appeared in *Frank Leslie's Illustrated Newspaper*, are attributed to Francis Collingwood, Jr. The Commission of Architects appointed to judge the designs was very pleased with Collingwood's plans and admired the monumental character and boldness of the design which, with its Florentine arches, harmonized well with the anchorages and the towers. The approach on the Brooklyn side had crossed all streets with iron bridges. On the New York side only Franklin Street was crossed by an iron bridge—all the other streets were spanned by masonry arches. Today, the approaches are lost in a tangled network of ramps, exits and entrances which connect the bridge and the F.D.R. Drive in Manhattan and Cadman Plaza in Brooklyn. **108. New York approach, ca. 1878.** Work on the cables and the approaches proceeded together. Here is the New York approach with the el station at Franklin Square in the background and beyond that the New York anchorage and tower.

When John A. Roebling presented his proposal for the bridge, he maintained that the approaches made of granite and limestone would "greatly beautify and improve this part of the city which appears to need it more than any other." The bridge approach in New York ran along the south line of the 4th Ward, one of the most impoverished and crime-ridden districts in the city. **109. New York approach.** The visual impact of the approaches on surrounding neighborhoods was one of the aspects of the bridge most praised by Montgomery Schuyler in his famous article in *Harper's Weekly* on May 26, 1883: "The street bridges are uniformly imposing by size and span and especially attractive also by reason of the fact that through them we get what is to be got nowhere else in our rectangular city, glimpses and 'bits' of buildings. The most successful of them all, and the most successful feature architecturally of all the masonry of the bridge, is the simple, massive, and low bridge of two arches which spans North William Street in New York."

Anchorage. York Street (triple arch). Main Street. Sands Street.

THE APPROACH TO THE BRIDGE FROM THE BROOKLYN SIDE.

Vandewater Street. Cliff Street. Franklin Square.

VIEW OF PORTION OF THE APPROACH ON THE NEW YORK SIDE.

North William Street. William Street. Rose Street.

PORTION OF THE APPROACH ON THE NEW YORK SIDE.

107

109

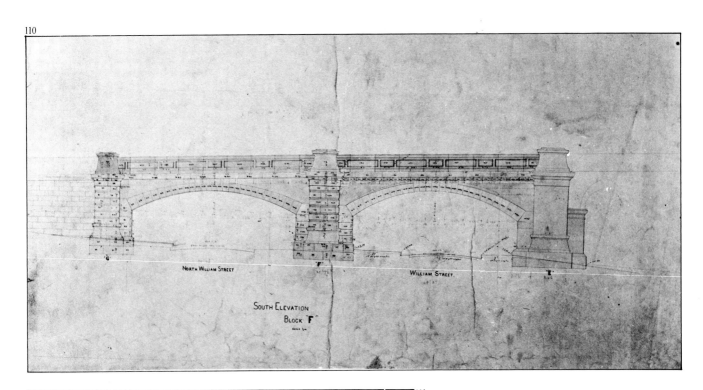

NORTH WILLIAM STREET

WILLIAM STREET

SOUTH ELEVATION
BLOCK "F"

77

3280

110. Arches over North William Street and William Street. These original engineering drawings are now in the Municipal Archives. Today the beautiful low arches so praised by Montgomery Schuyler no longer exist. They were replaced by plain steel-beam spans when the bridge was reconstructed to accommodate automobile traffic. **111. Design for iron fronts.** J. B. & J. M. Cornell, a New York foundry, submitted this oversize and beautifully rendered painting to obtain a contract for work to fill the arches and spaces between piers on the New York approach. **112. Elevation of storefronts.** This graceful row of arches was located just east of the William Street bridge. **113. Elevation and detail of the main arches, Brooklyn approach.** Individually rendered granite blocks, each with its own specifications, made up the row of arches behind the Brooklyn anchorage. Although the measurements are precisely and carefully calculated and rendered, the construction of these arches was marred by a fatal accident. In December 1877, one of them collapsed on top of a workman, crushing him so severely that he was scarcely recognizable. Upon inspection, it was learned that the wooden supports under the arch had been removed before the mortar had adequately dried and set. The engineer in charge was George McNulty.

ELEVATION OF STORE FRONTS
BLOCK "E"
NORTH.

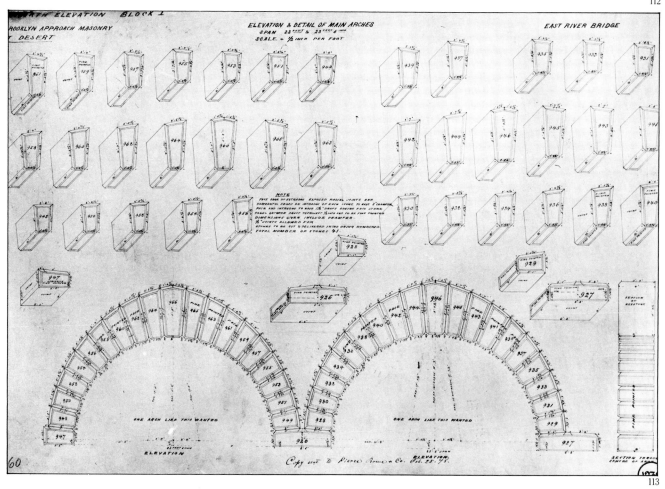

ARCH ELEVATION BLOCK I
BROOKLYN APPROACH MASONRY
T DESERT

ELEVATION & DETAIL OF MAIN ARCHES
SPAN 23 FEET & 23 FEET 6
SCALE = ½ INCH PER FOOT

EAST RIVER BRIDGE

ONE ARCH LIKE THIS WANTED

ONE ARCH LIKE THIS WANTED

ELEVATION

ELEVATION

Copy sent to Pierce Bros & Co. Feb. 25 75.

60

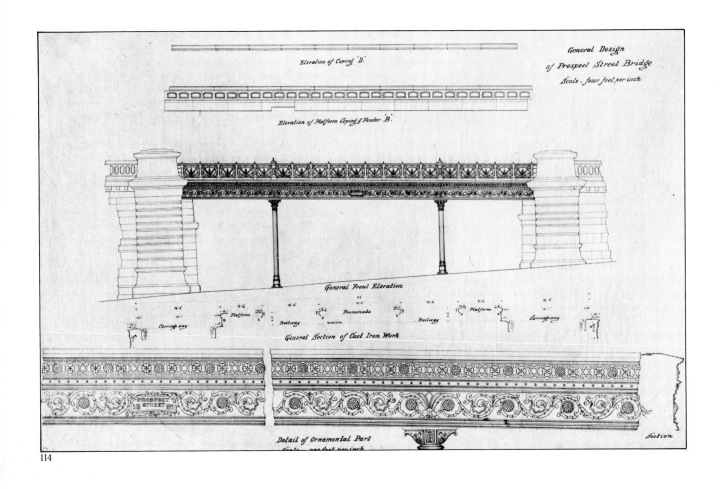

Elevation of Coping 'D'.

Elevation of Platform Coping of Fender 'B'.

General Front Elevation

Carriage way Platform Railway Promenade Railway Platform Carriageway

General Section of Cast Iron Work

Detail of Ornamental Part

Section

114

114. Prospect Street bridge. Another casualty of modernization, this elaborate and intricately detailed cast-iron bridge (it can also be seen in the right foreground of fig. 128) has been replaced by a stronger, more functional steel-beam structure. **115. Fulton Street, Brooklyn, 1880.** As soon as the cables were wrapped, the suspenders could be attached. Here the suspenders can be seen, very faintly, dangling in the breeze. Fulton Street in 1880 appears very much as it does today. All of the buildings shown on the north side of the street were erected between 1835 and 1837 and, with the exception of the building with the striped awnings, all have survived to this day. The Fulton Ferry House, which was completed in 1871 just as construction of the Brooklyn Bridge was getting under way, was torn down in 1926, two years after ferry service had been terminated.

116

116. Panorama of Brooklyn from the New York tower, 1880.
The cables are completed and the process of hanging the suspenders has just begun. The Brooklyn anchorage can be seen through the tower's Gothic double arches, and at the foot of the tower is the familiar Fulton Ferry Terminal. The tall spire of Holy Trinity Church on the corner of Clinton and Montague Streets is clearly visible in the background, at the right. **117. Suspending floor beams.** This precise woodcut from *Scientific American* shows how the suspenders were attached to the cables. First, wrought-iron suspender bands, 5″ wide and ⅝″ thick, were fitted over the cables and closed by a screwbolt 1¾″ in diameter. This bolt also held the wire-rope

suspender. Each suspender was made of steel-wire rope from 1⅝″ to 1¾″ in diameter and was capable of holding 50 tons or more. **118. The floor beams, from below.** On the lower end of each suspender, a threaded cast-iron socket held a stirrup rod which in turn held the floor beam. Since it was impossible to cut and fasten the suspenders to the exact length needed for the slightly inclined grade, the stirrup rods were made with long screw threads so that each beam could be raised or lowered to provide the desired grade. The floor beams were made of steel, weighed four tons each, measured 32″ deep and 9⅜″ wide and were 85′ long, the width of the bridge.

117

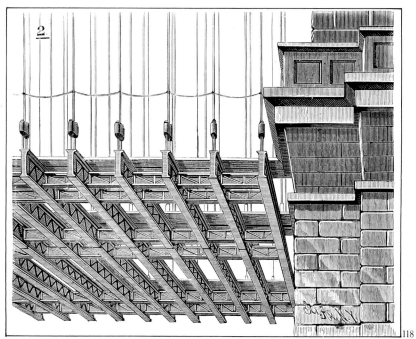

2

118

119

119. Suspending the floor beams, 1881. C. C. Martin stands at the far right with William H. Paine next to him. The other men, unidentified, are probably members of the engineering staff. Under their supervision, the work of suspending the floor beams proceeded uniformly from both sides of the towers, inland to the anchorages and outward to the center of the bridge, to avoid any lateral stress on the towers. After the floor beams were put in place, six longitudinal trusses were installed. They stiffened the bridge floor and provided a framework for the five roadways—the two outer divisions for horsecars and other vehicles, the two intermediate ones for a system of cable cars and a central raised platform for the pedestrian promenade. **120. Suspending the floor beams,**

1881. Another photograph shows two men, probably engineers, surveying the work in progress. At this time, the Edge Moor Iron Company was so late in its deliveries of steel beams that work on the bridge floor had to be stopped altogether. This resulted in the usual flurry of accusations of bribery and graft. Despite the aggravating delays, however, all the floor beams were in place by December 1, 1881. To celebrate this accomplishment, Emily Roebling, escorted by the Mayors of New York and Brooklyn, walked across the new floor (a 5'-wide wooden pathway had been laid on top of the steel floor beams) from Brooklyn to Manhattan. Under the New York tower they drank a champagne toast to the success of the bridge.

121

121. **"A Never-Ending 'Job.'"** After Emily and Washington Roebling returned to New York in 1876, Emily kept a scrapbook of all the articles about the bridge that appeared in the New York papers. The scrapbook, now a part of the Roebling Collections at Rensselaer Polytechnic Institute, includes this cartoon, which appeared in *Puck* on September 21, 1881, chastising the bridge directors for the delay in steel deliveries. The accompanying article says, "At present all labor seems to be at a stand-still because the commissioners . . . can not find the means to coerce a company that is trying to dodge its contract to supply the iron [sic] for the bridge. No one supposes that the commissioners, individually, would permit business associates to humbug them after this fashion; and it is not to be wondered at if an indignant public at last falls back upon a bad old pun, and asks indignantly if these supine gentlemen are waiting for more 'steel.'" **122. Removal of the footbridge, 1882.** By next spring, with all of the floor beams in place, a crew of 12 men and a foreman proceeded to remove the footbridge, no small task given that each supporting rope of the footbridge weighed 22 tons. Another change in the specifications of the bridge was announced around this time. One thousand tons of steel would be added to the construction of the bridge floor to make it strong enough to carry heavy locomotives. The Brooklyn trustees dreamed of the day when the New York Central would terminate in Brooklyn via the Great Bridge. As Roebling put it, they wanted to go from Brooklyn to their summer homes in Saratoga without having to change trains.

122

123

123. View from the Brooklyn tower, 1882. The floor beams are completely installed and the footbridge has been removed in this photograph, which shows the Brooklyn approach curving back toward Sands Street. This view is especially intriguing because of the angle from which it was shot. The photographer was probably standing on a temporary stairway on the left side of the Brooklyn tower halfway between the roadway and the top of the tower.

124

124. Installation of the trusswork, 1882. The photograph of the bridge, taken from downstream, shows the trusswork surrounding the cable-car system being installed. Most of the major work on the bridge was finished, but because of more delays in steel deliveries, the bridge would not be completed until the spring of 1883. Incredibly, these latest delays stirred up a movement on the Board of Trustees to retire Roebling and appoint C. C. Martin to take his place as Chief Engineer. The composition of the Board of Trustees had changed considerably since 1869. More than half its members were newcomers and had never seen, let alone spoken to, Roebling. Goaded by allegations in the press of political chicanery, the board members were impatient to get the job done. Thus, when the newest delay was announced, Seth Low, the 32-year-old Mayor of Brooklyn, started a movement to unseat the Chief Engineer. His attempts, however, failed and in September 1882, the Board narrowly voted down Low's resolution. Roebling remained Chief Engineer. **125. New York tower, 1883.** The scattered dots clinging to the weblike superstructure are workmen lashing the suspenders to the diagonal stays. The bridge is very nearly completed, the only details that date this photograph being the spiral staircase on the far left and the two boom-derricks on top of the tower. **126. Completing a great work, 1883.** A wood engraving shows a close-up of the workers far above the busy traffic on the East River. *Frank Leslie's Illustrated News*, which published this illustration, described their routine: "Each rigger is provided with a large pail of marlin, or tarred line, with which he lashes the stays and suspenders together until the little diamond shaped spaces are perfect in form. This is done so that when the superstructure is lowered by means of the screws at the base of the suspenders the strain will be equally distributed. The marlin lashings are temporary and will shortly be replaced by permanent iron clamps."

126

125

127

127. The New York terminal under construction. 128. The Brooklyn terminal under construction. These early 1883 photographs show work under way on the New York and Brooklyn terminals, which would not be finished in time for the bridge's official opening on May 24, 1883. The terminals were perhaps the least successful elements in the entire bridge project. Highly ornate cast-iron structures painted a dark red, they were excoriated in Montgomery Schuyler's May 24, 1883 article in *Harper's Weekly* as "grossly illiterate and discreditable to the great work." Over the years, as bridge traffic grew, so did the terminals. The New York terminal eventually projected all the way across Park Row into City Hall Park, and the train storage yards at the Brooklyn terminal stretched from High Street to Tillary Street. (The Prospect Street bridge, shown in the photograph of the Brooklyn terminal, conforms precisely with the plans shown in Fig. 114.) **129. Gable end of the Brooklyn terminal.**

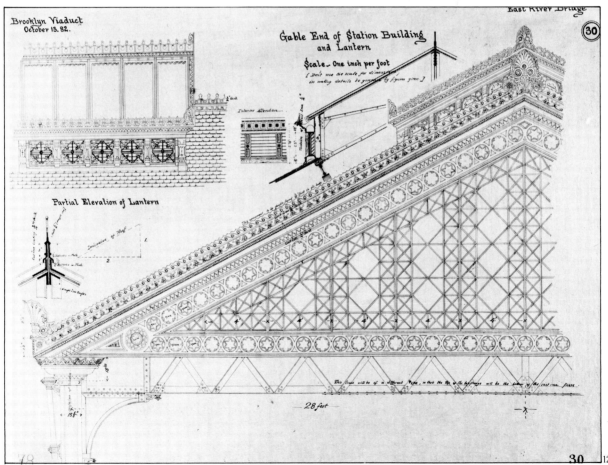

130

130. The Manhattan terminal. On May 24, 1883, the Brooklyn Bridge was officially opened. The front page of *Harper's Weekly* shows the complete Manhattan terminal with an inset of John A. Roebling, the creative genius whose greatest achievement was realized by the arduous and endlessly inventive efforts of his son. *Harper's* remarked, "The bridge is likely to outlast all the traditions of the men who built it. But one of the most enduring of these traditions is likely to be the touching story of the successive masters of the work, the Roeblings, father and son."

131

131. Washington Roebling. Roebling, who had watched the construction of the bridge from the window of his home on Columbia Heights, is the subject of this wood engraving which appeared on the cover of *Frank Leslie's Illustrated News*. Since he was too ill to take part in the official opening ceremonies on the bridge, Emily arranged a reception at their home which was attended by President Chester A. Arthur, Governor Grover Cleveland, the Bridge trustees, the engineers and over 1,000 other guests. *132 (Overleaf)*. **President Chester A. Arthur crosses the bridge.** The grand procession over the bridge, starting out from New York, was led by the Seventh Regiment, which, as it approached the Brooklyn tower, separated, stood at the sides of the promenade and presented

arms. President Chester A. Arthur, Governor Grover Cleveland, Mayor Edson of New York and other city officials proceeded through the regimental honor guard to the Brooklyn tower where they were met by Mayor Seth Low, the Bridge trustees and other important citizens of Brooklyn. *133 (Overleaf)*. **Mayor Seth Low welcomes the President to Brooklyn.** As Mayor Low greeted the President, a signal flag was dropped and the panoply of boats and ships, including the North Atlantic Squadron of the U.S. Navy, saluted with whistles, bells and cannon fire. A band played *Hail to the Chief*, celebrating, as the *New York Sun* observed, "the climax of fourteen years' suspense," the long-awaited completion of the Brooklyn Bridge.

THE SIGNAL TO FIRE THE PRESIDENTIAL SALUTE

MAYOR LOW RECEIVING THE PRESIDENT

COMIC BALLOONS

134

134. The Brooklyn terminal. The formal opening ceremonies took place in the Brooklyn terminal, which was twice the size of its counterpart in Manhattan. Six thousand people attended as William Kingsley officially presented the bridge to Mayors Edson and Low. Kingsley was the acting president of the Board of Trustees since Henry Cruse Murphy died in December 1882. With President Arthur and Governor Cleveland in attendance, the bridge trustees and countless city officials along with thousands of invited guests sat through three long hours of speeches praising the bridge and the men who had built it. **135. Invitation to the opening ceremonies, May 24, 1883.** The 6,000 invitations engraved by Tiffany's included a graceful rendering of the completed bridge. **136. "Representative Men in Religion, Literature, Science, Statesman-** **ship, Finance, Mercantile and Mechanical Progress, published in Commemoration of the Completion of the New York and Brooklyn Bridge. "** Not all the men in this composite photograph were actually involved in the building of the Brooklyn Bridge. A few of the familiar names are Wilhelm Hildenbrand (1), George McNulty (2), Washington A. Roebling (3), Francis Collingwood (4), C. C. Martin (5), W. H. Paine (6) and Sam Probasco (7) of the engineering class. Political figures include Henry Murphy (8), Seth Low (9), William Kingsley (10), James S. T. Stranahan (11; one of the earliest promoters of the bridge), Franklin Edson (12), Grover Cleveland (13), A. S. Hewitt (14), Chester A. Arthur (15), U. S. Grant (16), James A. Garfield (17) and Rutherford B. Hayes (18). Toward the front of this imposing gathering is John A. Roebling himself (19).

The Trustees of the
New York and Brooklyn Bridge
request the honor of the presence of
Mr. William Chittenden Lusk
at the
Opening Ceremonies
to take place on Thursday, May twenty fourth, at two o'clock, P. M.

Committees.

On behalf of the Board of Trustees. { William C. Kingsley, Pres. Henry W. Slocum. Jenkins Van Schaick.
James S. T. Stranahan. John T. Agnew. Otto Witte.

On behalf of the Cities, { Seth Low, Mayor of Brooklyn.
Franklin Edson, Mayor of New York.

On behalf of the Engineers. Washington A. Roebling.

138

137. Opening day, May 24, 1883. This rare photograph shows the bridge on opening day. The trustees had issued 7,000 tickets to select citizens who were allowed to cross the bridge before it was opened to the general public. All day long they strolled on the roadway as well as on the pedestrian promenade. At midnight the bridge would be opened to everyone. During the afternoon crowds had already started to gather around the terminals with many people jostling to be the first in line and first over the bridge. The Manhattan terminal can be seen through the left-hand arch of the

tower. **138. Souvenir medal.** Opening day meant a brisk business in the souvenir trade. Medals such as this, which people would wear on their lapels, were sold for 15 cents apiece. A telegraph office was open at each end of the bridge so people could send out telegrams to friends datelined from "Brooklyn Bridge." Sightseers could buy pamphlets, fans, hats, flags, scarves and models of the bridge. Some vendors were able to make a small fortune just by selling water and stale sandwiches. **139 (Overleaf). Fireworks and illumination, from the Brooklyn side.**

140. The great bridge at night. That evening there was a spectacular display of fireworks put on by Detwiller & Street, Pyrotechnics, a New York firm contracted by the Bridge Company. At 8:00 P.M., Miss Laura Detwiller applied the torch to the first flight of fifty rockets, the beginning salvo in a roaring display that lasted one hour. For the grand finale, 500 monster rockets were set off together with a thunderous explosion, producing a shower of millions of stars and golden rain which descended upon the bridge and river. As the last star disappeared in the water and the last spark died in the air, the whistles of the ferryboats saluted the new bridge, a salute which was taken up by all the other steamers in the river; for five minutes the air was filled with the din and screeching of whistles. As a final dramatic touch, according to the *New York Tribune*, "The

moon rose slowly over the Brooklyn tower and sent a broad beam like a benediction across the river." **141. The bridge in advertising.** At midnight, as the gates were opened to the public, the *New York Times* diligently recorded every "first" on the bridge: the first baby, the first beggar, the first drunken man (he collapsed opposite Franklin Square), the first love scene, etc. Advertisers also took up this mania to be first across the bridge. Thus, Dr. Scott's ad for electric hair brushes (not made of wire like the bridge but of pure bristles) tossed truth-in-advertising to the winds by claiming that his wagon was the "first and heaviest load to cross the Brooklyn Bridge. . . . President Arthur remarked that he would rather be Dr. Scott than President."

142. The panic of May 30, 1883. 143. The tragedy of the Brooklyn Bridge. 144 (*Overleaf*). Removing the bodies of the killed and injured. During the days following the opening of the bridge, large numbers of sightseers crowded onto the promenade, with several instances of pushing and shoving. On May 30, 1883, at about 4 P.M., when perhaps 20,000 people were on the bridge, a woman fell down the steps near the New York tower, another woman screamed and the crowd on the promenade pressed closer to see what was the matter. The momentum of the crowd caused others to fall down the stairs and within minutes there was a tremendous crush of human bodies so tightly packed together that some people bled through their noses and mouths. One policeman said that the

143

people at the sides of the promenade were crushed and scraped against the ironwork as they were helplessly pushed along by the frightened mob. The *New York Times* reported that the bewildered and struggling mass of men and women screamed, "For God's sake save us." Hats, canes, umbrellas and packages were thrown away. The women seemed helpless, and the men stood yelling and shouting, too bewildered to climb to safety on the girders. Children were saved by being passed hand to hand over the heads of the people within the crowd to rescuers located on the girders next to the cable-car tracks. As a tragic consequence of the disaster, 12 people died.

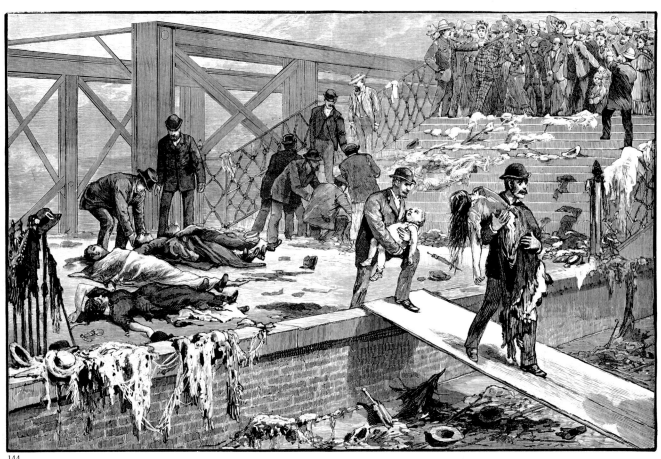

144

145 (Overleaf). **The bridge from Brooklyn Heights, 1883.** The best view of the bridge was from Brooklyn Heights, where one could see it rising majestically over the mansard tower of the Fulton Ferry House. In this wood engraving a ferryboat is leaving its Brooklyn slip while across the river another boat is on its way from New York. The Union Ferry receipts fell about 10 percent immediately following the opening of the bridge, but ferry officials remained sanguine about the new competition, remarking that a great deal of the traffic over the bridge was stimulated by curiosity: "Much of it will come back to us—the ferry boat offers a few minutes rest to man and beast. On the bridge it is a continual move on." *146.* **Manhattan view, 1883.** A view from the New York side shows an orderly procession of traffic. The cable cars are in operation, there is ample room on the promenade for pedestrians, and the roadway accommodates a steady flow of horses and wagons. The building on the far right, housing the *New York Tribune*, was completed in 1875. One of New York's tallest structures at that time, the Tribune Building accents the skyline of many of the photographs reproduced in this book. The five-story building left of it housed Tammany Hall until 1868, when it was sold to the *New York Sun*. The four buildings just to the right of the bridge entrance would soon be razed to make way for the New York World Building, which was completed in 1890.

147

147. Brooklyn Bridge electric station, 1883. 148. New York and Brooklyn Bridge Railroad, cable engines, 1884. 149. Rendering of the Brooklyn Bridge power station. The engine room and the engines which propelled the endless wire rope to which the cable cars were attached, and the plant of the U.S. Illuminating Company, which provided electricity to light the bridge (the first to have electric lights), were located beneath the Brooklyn approach. The boilers that provided the steam power for these engines were in a separate power station adjoining the approach, at the corner of Washington and Prospect Streets.

148

149

150

150. Cable cars, 1883. The cable cars went into operation on September 24, 1883. They were driven at a speed of about 10 mph by an endless wire cable which can be seen quite clearly between the rails of the right-hand track in this photograph. Roebling's design assured that the movement of the cable cars would not endanger the bridge's structural balance. As one car went up the grade of the bridge toward the center, another was acting as a counterbalance by going down grade. Starting with one car, the number of cable cars per train was gradually increased until three- or four-car trains were being run. The trip across the bridge cost five cents. **151. Brooklyn Bridge locomotive.** The trains were switched at the terminals by steam locomotives. In the beginning the cable trains were run only during the day. Not until 1885 did night service begin. Between the hours of 1 A.M. and 5 A.M., trains were pulled across the bridge by locomotives at intervals of 15 minutes. During these hours, the cable mechanism and power station were shut down to save money.

By 1885, the bridge trains were transporting 20 million passengers a year. **152. Aftermath of the Blizzard of 1888.** On March 12, 1881, winds averaged 48 mph with gusts as high as 84 mph. The anemometer on top of the Equitable Building at 120 Broadway clocked the wind velocity at 75 mph before being snapped from its perch and blown away. The bridge had to be closed to pedestrian traffic and the cable cars were taken out of service. Misgivings about the stability of the bridge (some had said it would be blown away by the first strong wind) were certainly laid to rest during the Blizzard of 1888: there was no recorded swaying of the bridge floor even at the height of the storm. That morning a huge ice floe 6″ thick floated down the Hudson River to the Battery where it was drawn up the East River by the turning tide. It lodged between Manhattan and Brooklyn at the foot of the bridge. Since the bridge was closed to traffic, people crossed the river over the ice floe instead, an ironic return to the way things were before the bridge was built.

151

152

153

156

155

154

153. Runaway horse, 1890. Horses crossing the bridge were often spooked by the rapid-transit system of cable cars. Safety gates were placed near the exits to keep the runaway horses from galloping into the streets. If the man at the gate was not fast enough closing it, a scene such as this one would occur. **154. The promenade, 1890s.** A walk on the promenade of the Brooklyn Bridge became one of New York's favorite pastimes. As John A. Roebling envisioned in his 1867 report to the Bridge Company, "The elevated promenade . . . will . . . allow people of leisure, and old and young invalids, to promenade over the bridge on fine days, in order to enjoy the beautiful views and the pure air. I need not state that in a crowded commercial city, such a promenade will be of incalculable value. Every stranger who visits the city will at least take one or two walks on this promenade, and the receipts of the Bridge Company from this source alone will be quite large." When the bridge first opened, admission to the promenade cost one cent, but in 1895 even this small toll was abolished. **155. Odlum's jump off the Brooklyn**

Bridge, 1885. The jumping craze actually started in 1882, before the bridge was even completed. Robert Donaldson planned to jump off the bridge on a bet but was frustrated by bridge officials who had him banned from the bridge premises. After three attempts, Donaldson finally gave up. The first man to succeed in this effort was Robert Odlum who, also on a bet, jumped off the bridge on May 19, 1885. He died shortly thereafter from severe internal hemorrhaging. **156. Steve Brodie.** On July 24, 1886, Steve Brodie, the most famous jumper of all, said that he had jumped off the bridge and produced witnesses to confirm his story. When John L. Sullivan's father was told about Brodie, he said, "Jumped off the bridge? Any damn fool can do that. I thought he jumped over it." Brodie was able to make a lucrative career out of his boast and even starred in a Bowery play in which he reenacted his leap from the bridge. By the 1890s, 20 people had committed suicide by jumping off the Brooklyn Bridge. By 1982 almost 50 people had jumped off the bridge; 33 of them were killed.

157

157. The bridge from the Fulton Ferry, Manhattan, 1898. This photograph shows how, 15 years after its completion, the bridge still dominated views of the city. In scale and mass, it was a harbinger of the direction the city's growth would take. **158. Curve at the Brooklyn terminal, 1898.** 1898 was a year of great changes on the bridge. First, because the cable cars proved unequal to the task of carrying the tremendous number of passengers their service attracted, electric trolleys were put into use. The new tracks for the trolleys were laid on the roadway, allowing horse and cart traffic only one lane instead of two. Also, on July 1, 1898, the King's County Elevated began service across the bridge on the cable-car tracks. To provide adequate service during rush hour, both the cable cars and

the elevated were run sharing the same tracks and operated by the same cable mechanism. At all other times, however, the elevated trains were run by electricity. Installation of the elevated lines meant that passengers could cross the bridge without changing trains or paying an extra fare. Later that July, there was a tremendous traffic jam in which trolleys and wagons were backed up from Brooklyn to Manhattan. Suddenly, the bridge sagged a few inches at two points, 250′ east of the Manhattan tower and 250′ west of the Brooklyn tower. Roebling himself, along with the bridge engineers, gave the bridge a careful inspection and declared that all was well. They did not even recommend repairing the trusses that had buckled slightly during the incident.

CURVE AT BROOKLYN TERMINAL
NEW YORK & BROOKLYN BRIDGE.

COPYRIGHT 1898
GEO. P. HALL & SON.
PHOTOGRAPHERS.
NEW YORK

159

159. Trolley cars and elevated trains with Manhattan terminal in the background, ca. 1900. By 1907, five- and six-car trains were being run on a headway of one minute, making 60 trains per hour. In 1907 the count of traffic for a period of 24 hours showed a maximum of 265,636 persons. The Manhattan terminal had been enlarged several times to accommodate the growing number of passengers and by 1907, the terminal projected well beyond Park Row and into City Hall Park. **160. Manhattan terminal, 1907.** The footbridge in the background, in use since 1886, is being removed to make way for a new extension of the Manhattan terminal. **161. Curve at the Brooklyn terminal, 1908.** In 1908, the cable cars were discontinued and trolleys were run on both levels of the bridge. The tracks were heavily reinforced, and new latticework for the trolley wire was installed on the upper levels. Meanwhile, the Brooklyn terminal was enlarged with storage yards for the elevated trains that stretched from High Street to Tillary Street. In 1910, the Brooklyn Heights Association complained that the treatment of the Brooklyn terminal "has been a blight and well nigh destruction to the welfare and prosperity of a large area of Brooklyn." The Singer Tower, completed in 1907, the most prominent point on the Manhat-

tan skyline, was the world's tallest building at the time. **162 (Overleaf). Aerial view of the Brooklyn, Manhattan and Williamsburg Bridges, ca. 1920.** Traffic congestion on the Brooklyn Bridge was alleviated with the completion of the Williamsburg Bridge in 1903 and the Manhattan Bridge in 1909. During the following years, subway lines tunneling under the East River were constructed: in 1908, the Lexington Avenue line at South Ferry; 1919, the Seventh Avenue line at Old Slip; in 1920, the BMT at Whitehall Street. By 1920, despite the variety of passage across the East River, the Brooklyn Bridge was carrying up to eight times the daily load it had carried when it was first opened in 1883. In 1922, during a rush-hour traffic jam with a train stalled near the Manhattan tower, the cables above the north roadway slipped, the outside cable by 1¾" and the inside cable by ½". Roebling, however, assured everyone that there was no need to worry. "They say the big cable slipped," he responded to a reporter's query. "That's just what the designers and the constructors intended it to do. If the big cable had not slipped, one end of the bridge would have fallen down. There is no need of rebuilding the bridge. It will last for 100 or 200 years."

163

163. **Manhattan terminal, 1935.** *164.* **Manhattan entrance to the bridge, 1945.** *165.* **Transformation of the Brooklyn Bridge to a six-lane highway, 1954.** In 1944, the Manhattan terminal, which had grown to gargantuan proportions, was torn down, and the elevated trains were discontinued. On the Brooklyn side, too, the terminal and elevated storage yards were demolished to create S. Parkes Cadman Plaza. Trolley-car service survived for another six years but, in 1950, the bridge was once again closed to allow major construction on the roadway. In this last renovation, the trolley tracks were removed, along with the trusswork that separated the tracks from the roadway, creating space for two three-lane highways,

each 30′ wide. The roadbed of wooden blocks was replaced with a closed floor of the lightest possible type—steel mesh with a shallow concrete filling—which emits a perpetual high-pitched singing sound under the speeding wheels of bridge traffic. The entire reconstruction increased the total dead load of the bridge by about four percent, an acceptable amount according to the engineers, since the bridge traffic was then restricted to passenger cars. In Fig. 165 the intermediate trusswork is still in place on the north roadway while the south roadway has already been transformed into three traffic lanes.

166

166. The Brooklyn Bridge, May 24, 1973. The rain-soaked roadway was closed to automobile traffic and opened to pedestrians to commemorate the ninetieth anniversary of the opening of the Brooklyn Bridge. A modest number of people braved the weather to stroll across the bridge and take part in a fair which was held on Fulton Street in Brooklyn. **167. The Brooklyn Bridge, 1982.** When the bridge was completed in 1883, the Reverend Richard Storrs, who spoke at the opening ceremonies, called it "this alluring roadway, resting on towers which rise like those of ancient cathe-

drals: this lacework of threads interweaving their separate delicate strengths into the complex solidity of the whole." It rose gracefully over the rooftops of Brooklyn and Manhattan, miraculously spanning the East River and joining the two cities for the first time. It was the greatest engineering achievement of the nineteenth century, heralding the future and symbolizing New York's emergence as a city of the world. Now the Brooklyn Bridge, instead of soaring above the skyline, sits at the city's foot, against a solid background of stone, steel and glass—the towers of lower Manhattan.

CREDITS

(Abbreviations: FL = *Frank Leslie's Illustrated Newspaper;* HNMM = *Harper's New Monthly Magazine*; HW = *Harper's Weekly*; MCNY = Museum of the City of New York; NYCDH* = New-York City Department of Highways; NYCDT* = New York City Department of Transportation; NYHS = New York Historical Society; NYPL = New York Public Library; RPI* = Rensselaer Polytechnic Institute Archives; RU* = Rutgers University, Special Collections; SA = *Scientific American*; SI = Smithsonian Institution.

*Through error, these credits were omitted from the first printing of this book.

Introduction: FL, Nov. 23, 1872. **1:** HW. **2:** *Winter Scene in Brooklyn* (1817–20), oil painting by Francis Guy; MCNY. **3:** *View of Brooklyn, L.I., from U.S. Hotel, N.Y.* (detail); MCNY. **4:** NYPL. **5:** HW, Feb. 9, 1867. **6:** HW. **8:** redrawing from original by Mary Shapiro. **10–13:** RU. **15:** courtesy Blair Birdsall. **16:** RPI. **17, 18:** Buffalo and Erie County Historical Society. **19, 20:** courtesy Blair Birdsall. **21–24:** RPI. **25:** HW, May 11, 1872. **26–28:** RPI. **29:** NYCDT. **30:** RPI. **31:** RU. **32, 33:** courtesy Blair Birdsall. **34:** NYCDT. **35:** HNMM; courtesy Blair Birdsall. **36:** MCNY. **37:** NYCDT. **38–43:** FL, Oct. 15, 1870. **44:** HW, Dec. 17, 1870. **45, 46:** by Talfor; MCNY. **47:** Long Island Historical Society. **48:** by S. A. Holmes; MCNY. **49:** NYCDT. **50:** HW, Nov. 19, 1870. **51–55:** by S. A. Holmes; MCNY. **56:** FL, July 5, 1873. **57:** by Brady; MCNY. **58:** HNMM; courtesy Blair Birdsall. **59:** FL, May 24, 1873. **60:** FL, Aug. 30, 1873. **61:** HW, Nov. 1, 1873. **62, 63:** MCNY. **64:** Eastman House. **65:** FL, Nov. 20, 1875. **66:** by J. H. Beal; NYHS. **67, 68:** MCNY. **69:** RPI. **70:** HNMM. **72:** MCNY. **74–76:** SA. **77, 78:** SA, Aug. 4, 1877, NYHS. **79:** HNMM, May 1883. **80:** SI. **81, 82:** SA. **83:** SA, Supplement No. 48, Nov. 25, 1876; courtesy Blair Birdsall. **84:** Long Island Historical Society. **85:** by G. W. Pach; MCNY. **86:** *The New York Illustrated Times*, June 29, 1878. **87–91:** SA, March 3, 1877; courtesy Blair Birdsall. **92:** by G. W. Pach; MCNY. **93:** HNMM; courtesy Blair Birdsall. **94:** HW, Nov. 24, 1877. **95–98:** SA, May 18, 1878. **99:** by J. A. LeRoy; MCNY. **100, 101:** SA, Nov. 9, 1878; courtesy Blair Birdsall. **102:** HNMM, May 1883; courtesy Blair Birdsall. **103, 104:** MCNY. **105:** by G. W. Pach; MCNY. **106:** MCNY. **107:** FL, Dec. 1, 1877; NYHS. **108:** MCNY. **109:** HW, May 26, 1883. **110–114:** NYCDH. **115:** MCNY. **116:** South Street Seaport Museum. **117, 118:** SA, May 21, 1881. **119:** NYHS. **120:** MCNY. **121:** *Puck*, Sept. 21, 1881; NYHS. **122:** FL, Apr. 1, 1882. **123:** Eastman House. **124:** NYPL. **125:** South Street Seaport Museum. **126:** FL, Apr. 28, 1883. **127, 128:** MCNY. **129:** NYCDH. **130:** HW, May 26, 1883. **131:** FL, May 26, 1883. **132:** HW, June 2, 1883. **133:** FL, June 2, 1883. **134:** HW, May 26, 1883. **135:** MCNY. **136:** Library of Congress. **137:** courtesy Blair Birdsall. **138:** MCNY. **139:** HW, June 2, 1883. **140:** FL, May 26, 1883. **141:** HW, June 23, 1883. **142:** FL, June 9, 1883. **143:** HW, June 9, 1883. **144:** FL, June 9, 1883. **145:** HW, May 24, 1883. **146:** R. Schwarz, artist; Shugg Bros., photoengravers; Burrow-Giles Litho. Co., printers; MCNY. **147, 148:** MCNY. **149:** NYCDH. **150:** courtesy Blair Birdsall. **151:** courtesy Hugh Dunne. **152:** NYHS. **153, 154:** MCNY. **155:** FL, May 30, 1885. **156:** MCNY. **157, 158:** NYHS. **159:** courtesy Hugh Dunne. **160:** MCNY. **161:** Long Island Historical Society. **162:** NYPL. **163–165:** courtesy Hugh Dunne. **166:** by Peter Harris. **167:** by Mary Shapiro.